# Biofuels from Food Waste

# Biofuels from Food Waste
## Applications of Saccharification Using Fungal Solid State Fermentation

Antoine P. Trzcinski

CRC Press
Taylor & Francis Group
Boca Raton London New York

CRC Press is an imprint of the
Taylor & Francis Group, an **informa** business

CRC Press
Taylor & Francis Group
6000 Broken Sound Parkway NW, Suite 300
Boca Raton, FL 33487-2742

### Library of Congress Cataloging-in-Publication Data

Names: Trzcinski, Antoine Prandota, author.
Title: Biofuels from food waste : applications of saccharification using
fungal solid state fermentation / Antoine Prandota Trzcinski.
Description: Boca Raton : CRC Press, 2017. | Includes bibliographical references.
Identifiers: LCCN 2017014056 | ISBN 9781138093720 (acid-free paper)
Subjects: LCSH: Biomass energy. | Food waste. | Fungi--Biotechnologoy.
Classification: LCC TP339 .T78 2017 | DDC 662/.88--dc23
LC record available at https://lccn.loc.gov/2017014056

Visit the Taylor & Francis Web site at
http://www.taylorandfrancis.com

and the CRC Press Web site at
http://www.crcpress.com

# Contents

# Author

**Antoine P. Trzcinski, PhD,** received a PhD from the Chemical Engineering Department of Imperial College London. Dr. Trzcinski developed a novel process for producing biogas from municipal solid waste and for the treatment of landfill leachate. In 2009, he carried out research and development at a pilot scale on the production of value-added products from algae at Manchester University. He worked extensively on liquid and solid state fermentation processes using biomass such as sugarcane bagasse, rapeseed meal, coffee waste, waste glycerol, wheat bran, and soybean residue for the production of sugars, ethanol, enzymes, and biodiesel using integrated biorefinery concepts. His research interests include fouling mitigation in membrane bioreactors, characterization of soluble microbial products, identification of bacterial and archaeal strains, pharmaceutical and antibiotic removal from wastewater, fate of nanoparticles in the environment, and bioelectro stimulation of microbes to improve bioprocesses through interspecies electron transfer (IET). In 2016, he joined the University of Southern Queensland in Australia as a lecturer and teaches environmental engineering, environmental engineering practice, hydraulics, solid and liquid waste treatment, and applied chemistry and microbiology as well as continuing his research in these fields.

# 1 Bioconversion of Food Wastes to Energy

## 1.1 INTRODUCTION

Food waste (FW) is organic waste discharged from various sources including food processing plants, and domestic and commercial kitchens, cafeterias, and restaurants. According to FAO (2012), nearly 1.3 billion tonnes of foods including fresh vegetables, fruits, meat, and bakery and dairy products are lost along the food supply chain. The amount of FW has been projected to increase in the next 25 years due to economic and population growth, mainly in Asian countries. For example, the annual amount of urban FW in Asian countries could rise from 278 to 416 million tonnes from 2005 to 2025 (Melikoglu et al., 2013b). Typical foods wasted in Asia–Pacific countries and around the world are summarized in Table 1.1 (FAO, 2012).

FW is traditionally incinerated with other combustible municipal wastes for generation of heat or energy. It should be recognized that FW contains high levels of moisture, which may lead to the production of dioxins during its combustion together with other wastes of low humidity and high calorific value (Katami et al., 2004). In addition, incineration of FW can potentially cause air pollution and loss of chemical values of FW. These suggest that an appropriate management of FW is strongly needed (Ma et al., 2009a). FW is mainly composed of carbohydrate polymers (starch, cellulose, and hemicelluloses), lignin, proteins, lipids, organic acids, and a remaining, smaller inorganic part (Table 1.2). Hydrolysis of carbohydrate in FW may result in the breakage of glycoside bonds with releasing polysaccharides as oligosaccharides and monosaccharides, which are more amenable to fermentation. Total sugar and protein contents in FW are in the range of 35.5%–69% and 3.9%–21.9%, respectively. As such, FW has been used as the sole microbial feedstock for the development of various kinds of value-added bioproducts, including methane, hydrogen, ethanol, enzymes, organic acid, biopolymers, and bioplastics (Han and Shin, 2004; Rao and Singh, 2004; Wang et al., 2005a; Sakai et al., 2006; Yang et al., 2006; Ohkouchi and Inoue, 2007; Pan et al., 2008; Koike et al., 2009; He et al., 2012b; Zhang et al., 2013b). Fuel applications ($200–400/ton biomass) usually create more value compared to generating electricity ($60–150/ton biomass) and animal feed ($70–200/ton biomass). Due to inherent chemical complexity, FW also can be utilized for production of high-value materials, such as organic acids, biodegradable plastics, and enzymes ($1000/ton biomass) (Sanders et al., 2007). However, it should be noted that the market demand for such chemicals is much smaller than that for biofuels (Tuck et al., 2012). Therefore, this chapter intends to review the FW valorization techniques that have been developed for the production of various biofuels, such as ethanol, hydrogen, methane, and biodiesel.

**TABLE 1.1**
**Typical Wasted Foods in Several Asia–Pacific Countries and around the Globe**

| Waste (KT) | World | Asia | Southeastern Asia | Australia | Cambodia | China | Indonesia | Japan | Malaysia | New Zealand | North Korea | Philippines | South Korea | Thailand | Vietnam |
|---|---|---|---|---|---|---|---|---|---|---|---|---|---|---|---|
| Cereal | 95,245 | 52,374 | 12,599 | 1,380 | 506.1 | 18,990 | 4.588 | 413.4 | 183.4 | 28.6 | 253 | 215.7 | 628.4 | 1,999 | 2,706 |
| Rice | 26,738 | 22,668 | 10,792 | 0.4 | 506.0 | 6,046 | 3,307 | 139.4 | 50.2 | ND | ND | 162.7 | 458.2 | 1,997 | 2,478 |
| Sugar | 459.9 | 188.9 | 151.7 | 93.6 | ND | 0.4 | ND | 20.8 | ND | ND | ND | ND | ND | 151.7 | ND |
| Pulses | 2,735 | 1,134 | 241.6 | 36.0 | 0.9 | 142.3 | 38.0 | 7.1 | ND | 1.2 | 10.3 | ND | 2.0 | 7.0 | 8.6 |
| Oil crops | 18,424 | 13,590 | 2,515 | 3.9 | 3.8 | 9,017 | 2,238 | 69.6 | 1.4 | 0.1 | 15.2 | ND | 12.7 | 159.4 | 30.5 |
| Vegetable oil | 616.1 | 269.3 | 116.9 | ND | ND | 133.4 | ND | 13.0 | 116.9 | ND | ND | ND | ND | ND | ND |
| Vegetables | 81,441 | 59,949 | 2,710 | 54.1 | 46,9 | 39,286 | 755.0 | 1,224 | 64.8 | 73.2 | 414.2 | 242.5 | 1,555 | 339.5 | 777.2 |
| Beans | 1,049 | 447.3 | 218.1 | 1.1 | 0.9 | 49.1 | 37.2 | 6.5 | ND | 0.2 | 10.3 | 2.2 | 1.6 | 3.7 | 5.2 |
| Onions | 5,891 | 3,877 | 186.0 | 14.6 | ND | 2,107 | 99.9 | 68.1 | ND | ND | 3.5 | 6.9 | 139.5 | 5.5 | 22.7 |
| Peas | 412.7 | 145.1 | 2.1 | 7.2 | ND | 39.9 | ND | 0.4 | ND | 1.1 | ND | 0.3 | 0.1 | 0.1 | ND |
| Tomatoes | 12,874 | 7,415 | 104.2 | ND | ND | 3,181 | 85.3 | 100.7 | 1.6 | 9.5 | 8.3 | 9.9 | 57.6 | 7.3 | ND |
| Potatoes | 62,229 | 12,912 | 466.1 | 23.6 | ND | 7,501 | 250.0 | 177 | ND | 10.9 | 156.0 | 34.4 | 95.3 | 9.0 | 83.3 |
| Fruits | 53,796 | 28,328 | 4,529 | 30.9 | 30.5 | 8,323 | 2,706 | 749 | 89.1 | 43.4 | 153.5 | 1,183 | 276.6 | 786.4 | 531.0 |
| Apples | 5,742 | 4,116 | 13.2 | 5.9 | ND | 3,192 | 3.1 | 84.6 | ND | 22.4 | 72.8 | 3.8 | 49.0 | 1.2 | 5.1 |
| Bananas | 13,532 | 8,544 | 1,896 | 5.4 | 7.8 | 949.3 | 637.4 | 213.0 | 56.1 | 7.6 | ND | 901.3 | ND | 153.7 | 137 |
| Coconuts | 3,038 | 2,488 | 2,159 | ND | ND | 20.5 | 2,066 | ND | 1.3 | ND | ND | 7.8 | ND | 69.1 | 0.9 |
| Pineapples | 1,829 | 579 | 431.9 | ND | 2.2 | 97.7 | ND | 15.4 | ND | 0.3 | ND | 109.9 | 2.8 | 189.5 | 50 |
| Coffee | 105.0 | 33.3 | 28.3 | ND | ND | 0.033 | 20.9 | ND | 0.6 | ND | ND | 6.4 | ND | ND | ND |
| Milk | 16,560 | 10,887 | 183.3 | ND | 1.6 | 1,447 | 45 | ND | 3.8 | 164.8 | 4.9 | ND | 42.4 | 25.2 | 9.5 |

*(Continued)*

**TABLE 1.1 (Continued)**

**Typical Wasted Foods in Several Asia–Pacific Countries and around the Globe**

| Waste (KT) | World | Asia | Southeastern Asia | Australia | Cambodia | China | Indonesia | Japan | Malaysia | New Zealand | North Korea | Philippines | South Korea | Thailand | Vietnam |
|---|---|---|---|---|---|---|---|---|---|---|---|---|---|---|---|
| Cream | 33.9 | 0.1 | ND | ND | ND | 0.1 | ND | ND | ND | ND | ND | ND | ND | ND | ND |
| Butter | 84.0 | 1.7 | ND | ND | ND | ND | ND | ND | ND | ND | ND | ND | ND | 23.1 | ND |
| Animal fats | 174.1 | 1.8 | ND | ND | ND | 0.1 | ND | ND | ND | ND | ND | ND | ND | ND | ND |
| Meat | 1,184 | 183.2 | ND | ND | ND | ND | ND | 107.2 | ND | ND | ND | ND | 107.2 | 23.1 | ND |
| Offal | 63.0 | 19.6 | ND | 8.7 | ND | ND | ND | ND | ND | ND | ND | ND | ND | ND | ND |
| Poultry meat | 97.5 | 61.2 | ND | ND | ND | ND | ND | 34.5 | ND | ND | ND | ND | ND | 23.1 | ND |
| Annual waste production per capita (T) | 0.184 | ND | 0.130 | 0.277 | 0.173 | 0.061 | 0.130 | 0.129 | 0.113 | 0.280 | 0.211 | 0.130 | 0.098 | 0.130 | 0.130 |
| Population (millions) | 7,067 | 4,175 | 610 | 22.9 | 14.5 | 1,354 | 237.6 | 127.5 | 29.6 | 4.5 | 24.6 | 92.3 | 50.0 | 65.9 | 88.8 |
| Total FW (MT) | 1300[a] | 278[b] | ≥79.3[a] | ≥6.34[a] | 2.50[c] | 82.80[d] | ≥30.90[a] | 16.40[d] | 3.36[a,e] | ≥1.25[a] | 5.19[d] | ≥12.00[a] | 4.91[d] | ≥8.6[a] | ≥11.55[a] |

*Source:* Reprinted from *Fuel*, 134, Uçkun, K. E. et al., Bioconversion of food waste to energy: A review, 389–399, Copyright 2014c, with permission from Elsevier.

*Note:* FW: food waste, T: ton, KT: kilotons, MT: million tonnes.

[a] Gustafsson et al. (2011)
[b] Melikoglu et al. (2013b)
[c] Seng (2010)
[d] OECD (2007)
[e] Noor et al. (2013)

**TABLE 1.2**
**Composition of Mixed Food Waste**

| Moisture | Total Solid | Volatile Solid | Total Sugar | Starch | Cellulose | Lipid | Protein | Ash | References |
|---|---|---|---|---|---|---|---|---|---|
| 79.5 | 20.5 | 95.0 | NA | NA | NA | NA | 21.9 | NA | Han and Shin (2004) |
| 84.1 | 15.9 | 15.2 | NA | NA | NA | NA | NA | NA | Kim et al. (2004) |
| 80.0 | 20.0 | 93.6 | NA | NA | NA | NA | NA | 1.3 | Kwon and Lee (2004) |
| 85.0 | 15.0 | 88.5 | NA | NA | 15.5 | 8.5 | 6.9 | 11.5 | Rao and Singh (2004) |
| 79.1 | 20.9 | 93.2 | NA | NA | NA | NA | NA | NA | Ramos et al. (2012) |
| 75.9 | 24.1 | NA | 42.3 | 29.3 | NA | NA | 3.9 | 1.3 | Ohkouchi and Inoue (2006) |
| 87.1 | 12.9 | 89.5 | NA | NA | NA | NA | NA | NA | Kim et al. (2008c) |
| 80.8 | 19.2 | 92.7 | NA | 15.6 | NA | NA | NA | NA | Pan et al. (2008) |
| 80.3 | 19.7 | 95.4 | 59.8 | NA | 1.6 | 15.7 | 21.8 | 1.9 | Tang et al. (2008) |
| 82.8 | 17.2 | 89.1 | 62.7 | 46.1 | 2.3 | 18.1 | 15.6 | NA | Wang et al. (2008a) |
| 75.2 | 24.8 | NA | 50.2 | 46.1 | NA | 18.1 | 15.6 | 2.3 | Wang et al. (2008b) |
| 85.7 | 14.3 | 98.2 | 42.3 | 28.3 | NA | NA | 17.8 | NA | Zhang et al. (2008) |
| 82.8 | 17.2 | 85.0 | 62.7 | 46.1 | 2.3 | 18.1 | 15.6 | NA | Ma et al. (2009a) |
| 61.3 | 38.7 | NA | 69.0 | NA | NA | 6.4 | 4.4 | 1.2 | Uncu and Cekmecelioglu (2011) |
| 4.4 | 35.6 | NA | NA | NA | NA | 8.8 | 4.5 | 1.8 | Cekmecelioglu and Uncu (2013) |
| 81.7 | 18.3 | 87.5 | 35.5 | NA | NA | 24.1 | 14.4 | NA | He et al. (2012a) |
| 81.5 | 18.5 | 94.1 | 55.0 | 24.0 | 16.9 | 14.0 | 16.9 | 5.9 | Vavouraki et al. (2012) |
| 81.9 | 14.3 | 98.2 | 48.3 | 42.3 | NA | NA | 17.8 | NA | Zhang and Jahng (2012) |

*Source:* Reprinted from *Fuel*, 134, Uçkun, K. E. et al., Bioconversion of food waste to energy: A review, 389–399, Copyright 2014c, with permission from Elsevier.

Total solid, total sugar, starch, cellulose, lipid, protein, and ash contents are given in wt% on the basis of dry weight. Volatile solid contents are given as the %VS ratio on total solid basis.

## 1.2 ETHANOL PRODUCTION

Recently, global demand for ethanol has increased due to its wide industrial applications. Ethanol is mainly used as a chemical feedstock to produce ethylene with a market demand of more than 140 million tonnes per year, a key material for further production of polyethylene and other plastics. As such, bioethanol produced from cheap feedstocks has gained interest (Lundgren and Hjertberg, 2010; International Renewable Energy Agency, 2013). Traditionally, bioethanol is produced from cellulose and starch rich crops, for example, potato, rice, and sugar cane (Thomsen et al., 2003). Starch can be easily converted to glucose by commercial enzymes and subsequently fermented to ethanol particularly by *Saccharomyces cerevisiae*. However, the hydrolysis of cellulose is more difficult. FW hydrolysis becomes much harder if large quantities of cellulosic feedstocks are present in FW. Use of abundant and cheap wastes such as lignocellulosic, municipal, and FW has been explored as alternative substrates for ethanol production (Kim and Dale, 2004; Jensen et al., 2011).

### 1.2.1 PRETREATMENTS

Harsh pretreatment may not be necessary during the conversion of FW to ethanol prior to enzymatic hydrolysis (Kumar et al., 1998; Tang et al., 2008). Instead, autoclave of FW before fermentation is often required for improving product yield and purity, but at the cost of energy and water consumption. It should be noted that thermal treatment may lead to partial degradation of sugars and other nutritional components, as well as side reactions (e.g., Maillard reactions) through which the amounts of useful sugars and amino acids are reduced (Sakai and Ezaki, 2006). Moreover, fresh and wet FW appear to be more effective than rewetted dried FW (Kim et al., 2005). This is mainly due to the decreased specific surface area of the dried substrate, resulting in a decrease in the reaction efficiency between the enzymes and substrate. Therefore, the utilization of FW without a drying pretreatment is preferred as long as microbial contamination is manageable. Without thermal sterilization, acidic condition is needed to prevent microbial contamination and putrefaction (Ye et al., 2008a; Koike et al., 2009). As such, acid-tolerant ethanol-producing microorganisms such as *Zymomonas mobilis* have been employed for the fermentation of FW (Tao et al., 2005; Wang et al., 2008a).

### 1.2.2 SACCHARIFICATION

The conversion efficiency of FW to ethanol depends on the extent of carbohydrate saccharification as yeast cells cannot ferment starch or cellulose directly into bioethanol (Tubb, 1986). A mixture of α-amylase, β-amylase, and glucoamylase of various origins is more effective for substrates with higher molecular weight. Pullulanase has also been added to the list of saccharifying enzymes recently (Tomasik and Horton, 2012). As a direct endo-acting debranching enzyme, pullulanase can specifically catalyze the hydrolysis of α-1,6-glucosidic linkages of branched polysaccharides (e.g., pullulan, dextrin, amylopectin, and related polymers), resulting in the release of linear oligosaccharides. Small fermentable sugars (e.g., maltose, amylose, glucose, maltose syrups, and fructose) can be produced in saccharification process, whereas cellulases and xylanases including endoglucanase, exoglucanase, β-glucosidase, and β-xylosidase, can also be employed to improve the hydrolysis of cereals for conversion of starches to glucose (Ducroo, 1987).

Table 1.3 shows the glucose and ethanol yields of different types of FW. The highest glucose concentration of about 65 g reducing sugar (RS)/100 g FW was obtained with α-amylase at a dose of 120 U/g dry substrate, glucoamylase (120 U/g dry substrate), cellulase (8 FPU/g dry substrate), and β-glucosidase (50 U/g dry substrate) (Cekmecelioglu and Uncu, 2013). In a study by Hong and Yoon (2011), a mixture of commercial enzymes consisting of α-amylase, glucoamylase, and protease resulted in 60 g RS/100 g FW.

### 1.2.3 Process Configurations

High glucose yield is achievable by increasing enzyme concentration and temperature at different solid loads, agitation speeds, and hydrolysis times in the saccharification processes (Sharma et al., 2007; Ado et al., 2009; Shen et al., 2009; Zhang et al., 2010). High glucose concentration may result in catabolite repression of the enzymes (Oberoi et al., 2011b). Therefore, fed-batch and simultaneous saccharification and fermentation methods have been developed for achieving high ethanol yield from FW (Ma et al., 2009b; Oberoi et al., 2011b). The fed-batch culture has been commonly employed for the production of high concentration reducing sugars which can be further fermented to ethanol (Ballesteros et al., 2002). Compared to batch culture, Yan et al. (2012b) found that saccharification and subsequent ethanol fermentation were both improved significantly using fed-batch configuration, for example, the glucose bioconversion yield reached 92% of its theoretical value. Alternatively, Ssf can be deployed to mitigate risk of catabolite repression. This combines enzymatic hydrolysis and ethanol fermentation into a single operation for keeping the concentration of enzymatically produced glucose at a low level so as to mitigate inhibition to enzymatic hydrolysis (Hari Krishna et al., 2001). This combined process can be performed in a single tank, with lower energy consumption, higher ethanol productivity, in a shorter processing time using less enzyme (Ballesteros et al., 2002). Optimization of fermentation conditions is vital for the success of the Ssf process as enzymes and fermenting microorganisms may have different optimum pH and temperatures. In a study by Hong and Yoon (2011), about 60 g RS and 36 g ethanol were produced from 100 g of FW in 48-h fermentation. Koike et al. (2009) also reported production of ethanol from non-diluted FW (garbage) in a continuous Ssf process with an ethanol productivity of 17.7 g/L h. Ma et al. (2009a) investigated the Ssf process using kitchen garbage by acid tolerant *Zymomonas mobilis* without any sterilization. Approximately 15.4 g sugar per 100 g of garbage and 0.49 g ethanol per grams sugar was obtained within 14 h, giving an ethanol yield of 10.08 g/L h.

### 1.2.4 Other Strategies to Improve Ethanol Yield

To improve ethanol productivity, various strategies have been explored, including use of strains with high ethanol tolerance (He et al., 2009; Wang et al., 2012)

**TABLE 1.3**

**Ethanol Production from Food Wastes**

| Waste | Method | Vessel Type | Pretreatment | Microorganism | Duration (h) | Y (g RS/100 g FW) | Y (g/g FW) | Y (g/g RS) | P (g/h) | References |
|---|---|---|---|---|---|---|---|---|---|---|
| Bakery waste | Simultaneous | 14 L fermenter | None | S. cerevisiae | 14 | 54 | 0.25 | 0.46 | NR | Kumar et al. (1998) |
| KW | Repeated batch Simultaneous | 1 L fermenter with 0.8 L working vol. | None | S. cerevisiae ATCC26602 | 264 | 12.3 | 0.06 | 0.5 | 3.7 | Ma et al. (2007) |
| Mandarin waste, banana peel | Simultaneous | 500 mL flask | Drying, steam explosion | S. cerevisiae Anr, Pachysolen tannophilus | 24 | 25.2 | 0.11 | 0.4 | NR | Sharma et al. (2007) |
| FW | Separate | 500 mL flask 100 mL working vol. | None | S. cerevisiae KA4 | 16 | 23.4 | 0.12 | 0.49 | NR | Kim et al. (2008c) |
| FW | Simultaneous | Flask with 100 g FW | None | S. cerevisiae | 48 | 11.25 | 0.08 | NR | NR | Ma et al. (2008) |
| KW | Separate Continuous | Tower shaped reactor, 0.45 L working vol. | LAB spraying | S. cerevisice strain KF-7 | 15 | 11.7 | 0.03 | 0.26 | 24 | Tang et al. (2008) |
| KW | Simultaneous | Flask with 100 g FW | None | S. cerevisiae | 67.6 | 34.8 | 0.23 | NR | NR | Wang et al. (2008c) |
| FW | Continuous Simultaneous | Fermenter with 4.3 kg FW | LAB spraying | S. cerevisiae KF7 | 25 | 36.4 | 0.09 | 0.24 | 17.7 | Koike et al. (2009) |
| FW | Simultaneous | 1 L fermenter with 0.8 L working vol. | None | S. cerevisiae KRM-1 | 48 | 8.9 | 0.06 | NR | 10.08 | Ma et al. (2009a) |
| KW | Repeated batch Simultaneous | 250 mL flask 150 mL working vol. | None | Zymomonas mobilis GZNS1 | 14 | 15.4 | 0.07 | 0.49 | 10.08 | Ma et al. (2008) |
| FW | Simultaneous | 250 mL flask 200 mL working vol. | None | S. cerevisiae | 48 | 60 | 0.36 | 0.22 | NR | Hong and Yoon (2011) |

*(Continued)*

**TABLE 1.3 (Continued)**
**Ethanol Production from Food Wastes**

| Waste | Method | Vessel Type | Pretreatment | Microorganism | Duration (h) | Y (g RS/100 g FW) | Y (g/g FW) | Y (g/g RS) | P (g/h) | References |
|---|---|---|---|---|---|---|---|---|---|---|
| FW | Separate | 5 L fermenter with working volume of 3L | None | S. cerevisiae | 24 | 27 | 0.16 | NR | 1.18 | Kim et al. (2011c) |
| FW | Synchronous Saccharification | Fermenter with 200 g FW | None | Saccharomyces italicus KJ | 352 | 12.5 | NR | NR | 2.24 | Li et al. (2011) |
| Mandarin waste | Simultaneous | 100 mL baffled flasks | Drying | S. cerevisiae | 15 | 52 | 0.34 | NR | 3.5 | Oberoi et al. (2011a) |
| Banana peels | Simultaneous | 100 mL baffled flasks | Drying | S. cerevisiae | 15 | 37.1 | 0.32 | 0.43 | 2.3 | Oberoi et al. (2011a) |
| KW | Separate | 250 mL flask 100 mL working vol. | None | S. cerevisiae | 96 | 50 | 0.2 | 0.39 | NR | Uncu and Cekmecelioglu (2011) |
| KW | Separate | 250 mL flask 100 mL working vol. | None | S. cerevisiae | 48 | 64.8 | 0.23 | 0.36 | NR | Cekmecelioglu and Uncu (2013) |
| Bread waste | Separate | 300 mL flask 80 g bread waste | Drying | S. cerevisiae Ethanol Red | 72 | 37 | 0.27 | NR | NR | Kawa-Rygielska et al. (2012) |
| KW | Separate (fb) | 500 mL flask 200 g FW | None | S. cerevisiae H058 | 48 | 29 | 0.14 | 0.47 | NR | Yan et al. (2012a) |

*Source:* Reprinted from *Fuel*, 134, Uçkun, K. E. et al., Bioconversion of food waste to energy: A review, 389–399, Copyright 2014c, with permission from Elsevier.

*Note:* NR: not reported, FW: food waste, KW: kitchen waste, RS: reducing sugar, Y: yield, P: productivity, Simultaneous: simultaneous saccharification fermentation, Separate: separate saccharification fermentation, fb: fed-batch.

and cell recycle through sedimentation or membrane retention (He et al., 2012a). Recombination of bioethanol-producing strains with the amylase-producing gene or the development of new strains with improved ethanol tolerance has also been reported (Li et al., 2011). However, stability of the recombinant gene has not yet been proven. Cell recycling has been known to significantly improve the performance of the continuous fermentation process (Wang and Lin, 2010).

## 1.2.5 LARGE-SCALE ETHANOL PRODUCTION FROM FOOD WASTE

Pilot and full scale plants for ethanol production from various wastes have been reported. The pilot study by Kumamoto University and Hitachi Zosen Company showed that 60 L of ethanol could be produced from 1 ton of municipal solid wastes, while the residual by-products could be further used for biogas production (Japan-for-Sustainability, 2013). In Finland, ST1 Biofuel built a network of seven ethanol plants converting various kinds of wastes to ethanol with a total annual capacity of 11 ML (Energy Enviro Finland, 2013; ST1, 2013). In Spain, citrus wastes have been converted to ethanol with a yield of 235 L/ton dry orange peel (BEST, 2013; Citrotechno, 2013). E-fuel developed a home ethanol system supported with micro-sensors to convert sugar/starch rich liquid wastes into ethanol for homeowners and small businesses (E-fuel, 2009). A theoretical estimate based on the data presented in Tables 1.1 and 1.3 suggests that 36.2, 126.8, and 593 TL (teraliters) of ethanol might eventually be produced annually in Southeast Asia, Asia, and the world, respectively.

## 1.3 HYDROGEN PRODUCTION

Hydrogen ($H_2$) is used as compressed gas and has a high energy yield (142.35 kJ/g). FW rich in carbohydrates is suitable for $H_2$ production. Table 1.4 summarizes the recent studies on $H_2$ production from FW. It can be seen that the hydrogen yields ranged from 0.9 mol $H_2$/mol hexose to 8.35 mol $H_2$/mol hexose (Patel et al., 2012). The factors such as the composition of FW, pretreatments, and process configurations may affect $H_2$ production.

### 1.3.1 SUBSTRATE COMPOSITION

Hydrogen production potential of carbohydrate-based waste was reported to be 20 times higher than that of fat-based and protein-based waste (Show et al., 2012). This was partially attributed to the consumption of hydrogen toward ammonium using nitrogen generated from protein biodegradation. Kim et al. (2010) reported that the $H_2$ yield was maintained at around 0.5 mol $H_2$/mol hexose at the C/N ratio lower than 20, while the $H_2$ yield was found to drop at higher C/N ratio because of the increased production of lactate, propionate, and valerate. The $H_2$ yield was significantly enhanced and reached to 0.9 mol $H_2$/mol hexose when the C/N ratio was balanced with an alkaline shock.

**TABLE 1.4**

**Hydrogen Production from Food Waste**

| Waste | Vessel Type | Pretreatment | Microorganism | Duration (day) | HRT (day) | OLR (kg VS/m³d) | OLR (kg COD/m³) | Y (mol/mol Hexose) | Y (mL/g VS) | P (g H₂/Lh) | References |
|---|---|---|---|---|---|---|---|---|---|---|---|
| FW | Leaching bed reactor with 3.8 L working vol. | None | HSSS | 7 | 5 | NR | NR | NR | 160 | NR | Han and Shin (2004) |
| FW with sludge | 415 mL bottle with 200 mL working vol. | None | HSSS | 3 | Batch | NA | NA | 0.9 | 67 | 9.9 | Kim et al. (2004) |
| FW | 715 mL bottle with 500 mL working vol. | None | Acidogenic culture from CSTR | 6 | Batch | NA | NA | 1.8 | 92 | 6.8 | Shin et al. (2004) |
| FW | Bioreactor with 3 L working vol. | None | Anaerobic SS | 5 | NR | 8 | NR | 2.2 | 125 | 3.8 | Shin and Youn (2005) |
| FW | Bioreactor with 3 L working vol. | None | Anaerobic SS | 60 | 5 | 3 | NR | 2.4 | NR | NR | Youn and Shin (2005) |
| FW | CSTR with 10 L working vol. | None | SS | 150 | 1.3 | 38.4 | 64.4 | NR | 283 | 19.9 | Chu et al. (2008) |
| FW | 1 L bioreactor with 500 mL working vol. | None | Anaerobic SS | 2 | Batch | NA | NA | NR | 57 | NR | Pan et al. (2008) |
| FW | 7.5 L bioreactor with 3 L working vol. | Heat pretreatment (90°C 20 min) | SS | 3 | Batch | NA | NA | 2.05 | 153.5 | 19.2 | Kim et al. (2008c) |

*(Continued)*

**TABLE 1.4 (Continued)**
**Hydrogen Production from Food Waste**

| Waste | Vessel Type | Pretreatment | Microorganism | Duration (day) | HRT (day) | OLR (kg VS/m³d) | OLR (kg COD/m³) | Y (mol/mol Hexose) | Y (mL/g VS) | P (g H₂/Lh) | References |
|---|---|---|---|---|---|---|---|---|---|---|---|
| FW | ASBR with 4.5 L working vol. | None | HSSS | NR | SRT: 5.25 HRT: 1.25 | NR | NR | 1.12 | 80.9 | 10.2 | Kim et al. (2008a) |
| KW | Bioreactor with 1 L working vol. | None | SS | 2 | Batch | NA | NA | NR | NR | 1.0 | Lee et al. (2008) |
| FW | Rotating drum with 200 L working vol. | None | None | 30 | 4 | 22.65 | NR | NR | 65 | NR | Wang and Zhao (2009) |
| Apple pomace | 150 mL bioreactor with 100 mL working vol. | Enzymatic pretreatment | HSSS | 2 | Batch | NA | NA | NR | 134 | NR | Wang et al. (2010a) |
| FW | CSTR 500 L working vol. | Heat pretreatment (100°C 30 min) | HSSS | 90 | 21 | NR | 12.3–71.3 | 1.82 | NR | NR | Lee et al. (2010a) |
| KW | CSTR with 20 L working vol. | None | SS | 59 | 4 | NR | NR | NR | NR | 7.1 | Lee et al. (2010b) |
| FW | SCR with 10 L working vol. | None | HSSS | 96 | 19 | NR | 39 | 2.5 | 114 | 41.3 | Lee et al. (2010b) |
| FW | ASBR with 0.15 m³ working vol. | Alkaline pretreatment (pH 12.5, 1 d) | HSSS | 200 | 36 | NR | NR | 0.9 | NR | NR | Kim et al. (2010) |

*(Continued)*

## TABLE 1.4 (*Continued*)
## Hydrogen Production from Food Waste

| Waste | Vessel Type | Pretreatment | Microorganism | Duration (day) | HRT (day) | OLR (kg VS/m³d) | OLR (kg COD/m³) | Y (mol/mol Hexose) | Y (mL/g VS) | P (g H₂/Lh) | References |
|---|---|---|---|---|---|---|---|---|---|---|---|
| FW | Bottle with 200 mL working vol. | Ultrasonication with acid | None | 14.6 | Batch | NA | NA | NR | 118 | NR | Elbeshbishy et al. (2011) |
| FW | Bottle with 200 mL working vol. | None | None | 3 | Batch | NA | NA | 1.79 | NR | 33.0 | Kim et al. (2011a) |
| FW | 500 mL bioreactor with 200 mL working vol. | None | HSSS | 1 | Batch | NA | NA | NR | NR | 6.6 | Mohd Yasin et al. (2011) |
| FW | 300 mL bioreactor with 150 mL working vol. | None | HSSS | 2 | Batch | NA | NA | NR | NR | NR | Ramos et al. (2012) |
| FW | Bioreactor with 150 mL working vol. | Lactate fermentation | Irradiated *R. sphaeroides* KD131 | 1 | Batch | NA | NA | 8.35 | NR | NR | Kim and Kim (2013) |

*Source:* Reprinted from *Fuel*, 134, Uçkun, K. E. et al., Bioconversion of food waste to energy: A review, 389–399, Copyright 2014c, with permission from Elsevier.

*Note:* FW: food waste, KW: kitchen waste, Y: yield, P: productivity. ASBR: anaerobic sequencing batch reactor, SBR: sequencing batch reactor, SS: seed sludge, HSSS: heat shocked seed sludge, NR: not reported, NA: not applicable.

## 1.3.2 PRETREATMENTS

Typically, mixed cultures have been employed for $H_2$ production from waste materials. However, hydrogen generated by *Clostridium* and *Enterobacter* is often readily consumed by hydrogenotrophic bacteria (Li and Fang, 2007). Seed biomass is generally pretreated with heat to suppress hydrogen consumers (Elbeshbishy et al., 2011). FW itself can be a source of $H_2$-producing microflora. Kim et al. (2008a) have applied several pretreatments to select microflora for hydrogen production. Lactic acid bacteria are the most abundant species in untreated FW, while $H_2$-producing bacteria are dominant in the pretreated FW. Heat treatment is effective for suppressing lactate production and increasing $H_2$/butyrate production. However, heat treatment is likely to increase costs in large-scale operations. Luo et al. (2010) investigated different pretreatment methods of inoculums, and concluded that pretreatment would only have short-term effects on hydrogen production, and the pretreatment is not very crucial (Wang and Zhao, 2009).

## 1.3.3 PROCESS CONFIGURATIONS

Various fermentation systems, such as the batch, semi-continuous, continuous, one or multiple stages, have been developed for production of $H_2$ from FW (Hallenbeck and Ghosh, 2009). High $H_2$ production rates have been reported in the anaerobic sequencing batch (ASBR) and up-flow anaerobic sludge blanket (UASB) reactors due to their high reactor biomass concentrations (Kim et al., 2008a). In these processes, the solid retention time (SRT) determines the substrate uptake efficiency, microbial size and composition, and metabolic pathway. A long SRT favors the growth of $H_2$ consumers, while a short SRT may reduce substrate uptake efficiency, active biomass retention, and subsequently the overall process efficiency. If the optimal SRT could be achieved at a low hydraulic retention time (HRT), it would enhance the productivity and technical feasibility of the $H_2$ production process (Wang and Zhao, 2009). Kim et al. (2008a) investigated the effects of SRT in the range of 24–160 h and HRT of 24–42 h on hydrogen production from FW. It was found that the maximum $H_2$ yield of 80.9 mL$H_2$/g volatile solid (VS), equivalent to 1.12 mol $H_2$/mol hexose, was obtained at SRT of 126 h and HRT of 33 h. Wang and Zhao (2009) obtained a hydrogen yield of 65 mL $H_2$/g VS at a long SRT of 160 d in a two-stage process. It is still debatable as for the effect of the organic loading rate (OLR) on bioconversion of FW to $H_2$. In some studies, lower $H_2$ yields were observed at higher OLRs, whereas the opposite trend was also reported in the literature. It appears that an optimal OLR would exist for the maximum $H_2$ yield (Wang and Zhao, 2009). Wang and Zhao (2009) reported that the hydrogen fermentation pathway became dominant and $H_2$ yield was steady at lower OLR ($\leq 22.65$ kg VS/m$^3$ d), while a decrease in the hydrolysis rate of substrate and an increase of propionic and lactic acids were observed. These suggest the possibility of co-production of organic acids if the cost related to separation is comparable with the value of the products. The inhibitory effect of organic acids produced at high OLR was also reported by Shin and Youn (2005).

Therefore, it is important to determine the optimum OLR and SRT for improving $H_2$ production. Acidity of the fermentation medium is another crucial parameter influencing the fermentation efficiency. It had been reported that the optimum pH for $H_2$ production from organic waste ranged from 4.5 to 6.5 (Kyazze et al., 2007). The accumulation of fermentation products, that is, $CO_2$, increases the acidity and then inhibits the microbial growth. Such fermentation products can be removed from the fermentation medium by simple gas sparging and mixing. The addition of alkaline or inoculum recycling are also frequently used for pH control (Li and Fang, 2007; Lee et al., 2008; Wang and Zhao, 2009; Kim et al., 2010; Lee et al., 2010a,b). Compared to the addition of alkali, sludge recirculation is an economically preferable approach for pH control. The long-term stability of a continuous two-stage process was maintained by recirculating high-alkalinity sludge, for example, at a OLR of 39 g COD/L d and HRT of 1.9 d, the system was stabilized at 2.5 mol $H_2$/ mole hexose, 114 mL $H_2$/g VS, and 462.5 mL $H_2$/L h over a period of 96 days (Lee and Chung, 2010).

The bioconversion yield of FW to $H_2$ production is low, for example, only about 33% of COD in organic materials can be harvested as $H_2$, while most of the energy content in the feedstock mainly end up as organic acids, such as acetic, lactic, and butyric acids. In other words, actual $H_2$ yield is much smaller than its theoretical value of 12 mol $H_2$/mol glucose (Kim and Kim, 2013). As a result, the commercial value of organic acids—particularly lactic acid—should be further explored. To improve economic viability of the bioconversion process, $H_2$ production should also be combined with the methane, organic acids, and ethanol production processes (Lin et al., 2013). Kyazze et al. (2007) reported that the efficiency of the $H_2$ production process was improved using the two-stage $H_2$-methane production process. Lee et al. (2010a) reported the feasibility of continuous $H_2$ and $CH_4$ fermentation in a two-stage process using sludge recirculation from the sludge storage tank (denitrification + digestion sludge storage) in a full-scale system. Even so, only 2.5 mol $H_2$/mol hexose was obtained due to the limitations of anaerobic metabolism.

Alternatively, photo-fermentation has also been explored for the conversion of organic acids to $H_2$. In order to increase the overall $H_2$ yield, a combined dark- and photo-fermentation system has been proposed. In this process, lactic acid produced from FW is utilized by photo-fermentative bacteria, particularly purple non-sulfur bacteria and finally converted to $H_2$ while the remaining residue is converted to $CH_4$ (Show et al., 2012). Overall, via the three-stage fermentation system, 41% and 37% of the energy content in the FW could be harvested as $H_2$ and $CH_4$, respectively, corresponding to the electrical energy yield of 1146 MJ/ton FW (Kim and Kim, 2013). Lee and Chung (2010) conducted a cost analysis of hydrogen production from FW using two-phase hydrogen/methane fermentation, and suggested that the abundance and low cost of FW makes it economically more feasible than the other sources for $H_2$ production. However, the economic feasibility of process applications from FW is dependent on the cost of FW collection. Besides, hydrogen production processes should be combined with an ancillary process, such as methane fermentation, to achieve complete treatment and disposal of FW. Last, it should also be recognized that the technological and economic challenges associated with the fermentative $H_2$ production and its purification, storage, and distribution may also slow down the wide application of bio $H_2$ as green energy.

## 1.4 METHANE PRODUCTION

The production of biogas, particularly methane via anaerobic processes, is an acceptable solution for waste management because of its low cost, low production of residual waste, and its utilization as a renewable energy source (Morita and Sasaki, 2012; Nasir et al., 2012). In addition to biogas, a produced nutrient-rich digestate can also be used as fertilizer or soil conditioner. Table 1.5 summarizes the studies pertaining to anaerobic digestion of various kinds of FW. Mtz. Viturtia et al. (1989) investigated two-stage anaerobic digestion of fruit and vegetable wastes, in which 95.1% volatile solids (VS) conversion with a methane yield of 530 mL/g VS was achieved. In a study by Lee et al. (1999), FW was converted to methane using a 5-L continuous digester fed with an OLR of 7.9 kg VS/m³ d, resulting in a 70% VS conversion with a methane yield of 440 mL/g VS. Gunaseelan (2004) has reported the methane production capacities of about 54 different fruit and vegetable wastes ranging from 180 to 732 mL/g VS depending on the origin of wastes.

Feedstock characteristics and process configuration are the main factors affecting the performance of anaerobic digestion (Molino et al., 2013). The physical and chemical characteristics of the waste, such as moisture, volatile solid, nutrient content, and particle size affect the biogas production and process stability. Cho et al. (1995) determined the methane yields of different FW over 28 days at 37°C, and found 482, 294, 277, and 472 mL/g VS for cooked meat, boiled rice, fresh cabbage, and mixed FW, with 82%, 72%, 73%, and 86% efficiency, respectively, based on elemental compositions of raw materials.

### 1.4.1 SINGLE-STAGE ANAEROBIC DIGESTION

The process configuration is very important for the efficiency of methane production process. Single-stage anaerobic digestion process has been widely employed for municipal solid waste treatment. As all of the reactions (hydrolysis, acidogenesis, acetogenesis, and methanogenesis) take place simultaneously in a single reactor, the system encounters less frequent technical failures and has a smaller investment cost (Forster-Carneiro et al., 2008). The anaerobic digestion can be wet or dry; the former uses the waste as received, while the latter needs to lower water content to about 12% of total solid (Nasir et al., 2012). Compared to wet anaerobic digestion, dry anaerobic digestion provides lower methane production and VS reduction due to the volatile fatty acid (VFA) transport limitation (Nagao et al., 2012). El-Mashad et al. (2008) reported that a digester treating FW was not stable due to the VFA accumulation and low pH, leading to low biogas production. On the other hand, the stability of single-stage anaerobic digester for easily degradable FW is of concern (Lee et al., 1999).

### 1.4.2 TWO-STAGE ANAEROBIC DIGESTION

In contrast to single-stage anaerobic digestion, two-stage anaerobic digestion has often been used for producing both hydrogen and methane in two separate reactors (Chu et al., 2008). In such a system, fast-growing acidogens and hydrogen-producing microorganisms are enriched for the production of hydrogen and volatile fatty acids

**TABLE 1.5**

**Methane Production from Food Wastes**

| Waste | Microorganism | Pretreatment | Process Type | Vessel Type | Duration (days) | HRT (days) | OLR (kg VS/m³d) | OLR (kg COD/m³d) | Biogas Yield (mL/g VS) | CH₄ Yield (mL/g VS) | %CH₄ | Efficiency (VS, %) | References |
|---|---|---|---|---|---|---|---|---|---|---|---|---|---|
| Fruit and vegetable waste | Cow manure | None | Two stage | Bioreactor with 0.5 L working vol. | 29 | 1 | 1–9 | NR | NR | 530 | 70 | 95.1 | Mtz. Viturtia et al. (1989) |
| FW | Anaerobic SS | Freeze drying of waste | Two stage | UASB with 8 L working vol. | 120 | NR | 1.04 | 7–9 | NR | 277–482 | NR | 90 | Cho et al. (1995) |
| FW | Anaerobic SS | None | Two stage | Continuous pilot scale 5 tons/d capacity | 90 | NR | 7.9 | NR | NR | 440 | 70 | 70 | Lee et al. (1999) |
| Fruit and vegetable waste | Anaerobic SS | None | Single stage | Serum bottles with 135 mL vol. | 100 | Batch | NA | NA | NR | 180–732 | NR | NR | Gunaseelan (2004) |
| FW and activated sludge | Anaerobic SS | None | Single stage | Semi continuous reactor with 3.5 L working vol. | 250 | 13 | 2.43 | 4.71 | NR | 321 | 64.4 | 55.8 | Heo et al. (2004) |

*(Continued)*

**TABLE 1.5 (Continued)**
**Methane Production from Food Wastes**

| Waste | Microorganism | Pretreatment | Process Type | Vessel Type | Duration (days) | HRT (days) | OLR (kg VS/m³d) | OLR (kg COD/m³d) | Biogas Yield (mL/g VS) | CH₄ Yield (mL/g VS) | %CH₄ | Efficiency (VS, %) | References |
|---|---|---|---|---|---|---|---|---|---|---|---|---|---|
| Potato waste | Anaerobic SS | None | Two stage | Packed bed with 1 L working vol. | 33 | NR | NR | 1–3 | NR | 390 | 82 | NR | Parawira et al. (2005) |
| FW | Anaerobic SS | None | Two stage | Bioreactor with 12 L working vol. | 60 | 20 | 8 | NR | NR | NR | 68.8 | 86.4 | Youn and Shin (2005) |
| FW | Bacteria isolated from landfill soil and cow manure | None | Single stage | 3 stage semi continuous with 8 L working vol. | 30 | 12 | NR | NR | NR | NR | 67.4 | NR | Kim et al. (2006b) |
| FW | Anaerobic SS | None | Single stage | Batch | 28 | 10–28 | NA | NA | 600 | 440 | 73 | 81 | Zhang et al. (2007) |
| FW | SS | None | Two stage | CSTR with 10 L working vol. | 150 | 5 | 6.6 | 16.3 | NR | 464 | 80 | 88 | Chu et al. (2008) |
| FW | Landfill soil and cow manure | None | Single stage | Batch 5 L | 60 | 20–60 | NR | NR | 490 | 220 | NR | NR | Forster-Carneiro et al. (2008) *(Continued)* |

**TABLE 1.5 (*Continued*)**
**Methane Production from Food Wastes**

| Waste | Microorganism | Pretreatment | Process Type | Vessel Type | Duration (days) | HRT (days) | OLR (kg VS/m³d) | OLR (kg COD/m³d) | Biogas Yield (mL/g VS) | CH$_4$ Yield (mL/g VS) | %CH$_4$ | Efficiency (VS, %) | References |
|---|---|---|---|---|---|---|---|---|---|---|---|---|---|
| FW | Bacteria and sludge from various sources | None | Three stage | UASB with 4800L working vol. | NR | 12 | 54.5 | ND | ND | 254 | 68 | 90.1 | Kim et al. (2008b) |
| FW | SS | None | Two stage | Bioreactor with 4.5 L working vol. | 200 | 1–27 | NR | 15 | 578 | 520 | 90 | NR | Park et al. (2008) |
| FW | SS | LAB pretreatment and SsF | Two stage | Bioreactor with 5 L working vol. | 98 | 7 | NR | NR | 850 | 434 | 51 | NR | Koike et al. (2009) |
| FW | No addition | None | Two stage | Rotating drum with 200 L working vol. | 30 | SRT 26.7 h | 4.61 | NR | 769 | 546 | 71.5 | 82.2 | Wang and Zhao (2009) |
| FW | SS | Heat pretreatment (100°C 30 min) | Two stage | UASB with 2.3 L working vol. | 60 | 3.9–6.4 | NR | NR | NR | NR | 80 | 80 | Lee et al. (2010a) |
| FW | SS | None | Two stage | Gas sparging type reactor with 40 L working vol. | 96 | 15.4 | NR | 4.16 | NR | NR | 65 | 88.1 | Lee and Chung (2010) |

*(Continued)*

**TABLE 1.5 (Continued)**
**Methane Production from Food Wastes**

| Waste | Microorganism | Pretreatment | Process Type | Vessel Type | Duration (days) | HRT (days) | OLR (kg VS/m³d) | OLR (kg COD/m³d) | Biogas Yield (mL/g VS) | CH₄ Yield (mL/g VS) | %CH₄ | Efficiency (VS, %) | References |
|---|---|---|---|---|---|---|---|---|---|---|---|---|---|
| FW | NR | None | Single stage | Digester with 900 m³ tank vol. | 426 | 80 | 2.5 | NR | 643 | 399 | 62 | 90 | Banks et al. (2011) |
| FW | Anaerobic SS | Enzymatic pretreatment | Two stage | UASB with 2.7 L working vol. | 75 | 2.2 | NR | 2.2 | NR | NR | 75 | 61 | Moon and Song (2011) |
| KW | Anaerobic SS | None | Two stage | Hydrolytic reactor (10 L), methanogenic MBR (3 L) | 19 | 23 | 10 | NA | NR | 357 | 63–70 | 81 | Trzcinski and Stuckey (2011) |
| FW | Anaerobic SS | Trace element addition | Single stage | Semi-continuous with 150 mL working vol. | 368 | 20–30 | 2.19–6.64 | NR | NR | 352–450 | 51.2 | NR | Zhang and Jahng (2012) |
| FW | Anaerobic SS | FW liquidized at 175°C for 1h | Single stage | UASB with 2 L working vol. | 72 | 4–10 | NR | 2–12.5 | NR | NR | 63 | 93.7 | Latif et al. (2012) |

(Continued)

**TABLE 1.5 (*Continued*)**
**Methane Production from Food Wastes**

| Waste | Microorganism | Pretreatment | Process Type | Vessel Type | Duration (days) | HRT (days) | OLR (kg VS/ m³d) | OLR (kg COD/ m³d) | Biogas Yield (mL/g VS) | CH₄ Yield (mL/g VS) | %CH₄ | Efficiency (VS, %) | References |
|---|---|---|---|---|---|---|---|---|---|---|---|---|---|
| FW | Anaerobic SS | None | Single stage | CSTR with 3 L working vol. | 225 | 16 | NR | 9.2 | NR | 455 | NR | 92.2 | Nagao et al. (2012) |
| FW and SS | Anaerobic SS | None | Single stage | Bioreactor with 6 L working vol. | NR | 8–30 | 4–21.8 | NR | 1039 | 465 | 52 | 90.3 | Dai et al. (2013) |
| FW | NR | None | Single stage | Digester with 800 mL working vol. | 30 | Batch | NA | NA | 621 | 410 | 66 | NR | Zhang et al. (2013b) |

*Source:* Reprinted from *Fuel*, 134, Uçkun, K. E. et al., Bioconversion of food waste to energy: A review, 389–399, Copyright 2014c, with permission from Elsevier.

*Note:* FW: food waste, KW: kitchen waste, SS: seed sludge, UASB: Upflow anaerobic sludge blanket reactor, SsF: simultaneous saccharification fermentation, MBR: membrane bioreactor, LAB: lactic acid bacteria. NR, not reported; NA, not applicable.

(VFAs) in the first stage. In the second stage, slow-growing acetogens and methanogens are built-up, and VFAs are converted to methane and carbon dioxide. In a study by Park et al. (2008), single-stage and two-stage thermophilic methane fermentation systems were operated using artificial kitchen waste. In both systems, the highest methane recovery yield of 90% (based on COD) was determined at the OLR of 15 g COD/L d. However, the propionate concentration in the single-stage reactor fluctuated largely and was higher than that in the two-stage process, indicating less stable digestion. Massanet-Nicolau et al. (2013) have also compared single- and two-stage anaerobic fermentation systems on FW processing. The methane yield in two-stage fermentation was improved by 37% and was operating at much shorter HRTs and higher loading rates. Lee and Chung (2010) also proved that two-stage hydrogen/methane fermentation has significantly greater potential for recovering energy than methane-only fermentation.

## 1.4.3 REACTOR CONFIGURATIONS

Packed bed reactors (PBRs) or fixed bed systems have been developed in order to attain high loading, immobilize microbial consortia, and stabilize methanogenesis (Kastner et al., 2012). Parawira et al. (2005) investigated the performances of two different systems, one consisting of a solid bed reactor for hydrolysis/acidification connected to an upflow anaerobic sludge blanket methanogenic reactor (UASB) while the other consists of a solid bed reactor connected to a methanogenic reactor packed with wheat straw as biofilm carriers (PBR) during mesophilic anaerobic digestion of solid potato waste. Although the PBR degraded the organic materials faster than the UASB, the methane yield (390 mL/g VS) and the cumulative methane production were equal in both systems. Among the high rate anaerobic reactors, the UASB reactor has been widely used to treat various kinds of organic wastes. The UASB provides the immobilization of anaerobic bacteria by granulation resulting in high microbial activity and good settling characteristics (Moon and Song, 2011). This also allows for high OLR and the maintenance of long retention time. Latif et al. (2012) investigated the mesophilic and thermophilic anaerobic treatment of liquidized FW in an UASB reactor by increasing the stepwise OLR and the temperature. The UASB reactor was efficient for COD removal (93.7%) and high methane production (0.912 L/g COD) due to low VFA accumulation under controlled temperature and pH. A temperature of 55°C and OLR of 12.5 g COD/L with 4 days HRT supported a maximum biogas production of 1.37 L/g COD. Continuously stirred tank reactors (CSTRs) and fluidized bed reactors (FBRs) were also investigated for methanogenesis (Kastner et al., 2012). Fermentation yielded 670 normalized liters (NL) biogas/kg VS with the CSTR and 550 NL biogas/kg VS with the FBR while the average methane concentration was approximately 60% for both reactor systems. However, the stability of the process was greater in the FBR. In summary, the two-stage process can attain higher OLR and higher methane generation. In addition, it is less vulnerable to fluctuations in OLR than a single methanogenic process. The efficiency of digestion could be improved by co-digesting different wastes, trace element addition, and using active inoculum as start-up seed. The highest methane yields from FW were reported by Koike et al. (2009). They obtained a biogas production of 850 mL/g VS during the two-stage hydrogen and methane production processing of FW. Approximately 85% of the energy of the garbage was converted to fuels, ethanol, and methane by this process. Considering the data in Tables 1.1 and 1.2 and the maximum methane yield of 546 mL/g VS reported in Table 1.5, it can be estimated that $1.32 \times 10^9$ m$^3$ methane can be produced annually which can generate $2.6 \times 10^7$ GJ energy using the total food waste generated in the world.

## 1.5  BIODIESEL PRODUCTION

FW can also be converted to fatty acids and biodiesel either by direct transesterification using alkaline or acid catalysts or by the transesterification of microbial oils produced by various oleaginous microorganisms (Chen et al., 2009; Mahmood and Hussain, 2010; Papanikolaou et al., 2011; Yaakob et al., 2013). Microbial oils can be produced by many yeast strains and they can be used as substitute for plant oils due to their similar fatty acid compositions. Alternatively, they can be used as raw material for biodiesel production (Uçkun et al., 2013b). Recent publications on the production of microbial lipids from various FW using different microbial strains are listed in Table 1.6. Pleissner et al. (2013) have revealed the potential of FW hydrolyzate as a culture medium and nutrient source in microalgae cultivation for biodiesel production. The FW hydrolyzate was prepared using *Aspergillus awamori* and *Aspergillus oryzae* and then used as a culture medium for the growth of heterotrophic microalgae *Schizochytrium mangrovei* and *Chlorella pyrenoidosa*. The microorganisms grew well on the FW hydrolysate leading to the production of 10–20 g biomass. The majority of fatty acids present in lipids of both strains was reported to be suitable for biodiesel production. Papanikolau et al. (2011) investigated the capacities of five *Aspergillus* sp. and *Penicillium expansum* to produce lipid-rich biomass from waste cooking olive oil in a carbon-limited culture. A significant amount of lipid accumulation was determined in each culture while the highest lipid yield (0.64 g/g dry cell weight) with a productivity of 0.74 g/g was obtained by *Aspergillus* sp. ATHUM 3482. The fatty acids accumulated were mainly C18:1 and they have the potential to develop food/feed supplements. From Table 1.6, it can be seen that the studies related to mixed food waste is still very scarce and that the productivity is relatively low. In addition, an extraction and a transesterification step are required to obtain biodiesel. The residual water in FW that is inhibitory in the transesterification is an additional obstacle for this type of fermentation from mixed food waste. Vegetable oils, butter, and animal fats that are produced globally were presented in Table 1.1. Assuming a maximum lipid yield of 0.74 g/g oil that was obtained from waste cooking oils and with a transesterification yield of 0.95 FAME/g lipid, it can be estimated that 647 kT (kilotons) of biodiesel can be produced annually in the world. This can potentially generate $24.5 \cdot 10^6$ GJ energy per year globally.

## 1.6  CONCLUSIONS

The management of FW poses a serious economic and environmental concern. It appears from this chapter that bioconversion of FW to energy in terms of ethanol, hydrogen, methane, and biodiesel is scientifically possible. However, difficulties

**TABLE 1.6**

**Fatty Acids and Biodiesel Production from Food Wastes**

| Waste | Microorganism | Pretreatment | Vessel Type | Conditions | Duration (days) | Y (g cell/ g waste) | Y (g lipid/g cell) | Y (g lipid/g fat consumed) | P (g polymer /Lh) | μ (h$^{-1}$) | References |
|---|---|---|---|---|---|---|---|---|---|---|---|
| Waste cooking olive oil | *A. niger* NRRL363 | Filtration | SmF-250 mL flasks | 28°C, pH 6, 200 rpm | 5 | 1.2 | 0.49 | 0.6 | NR | NR | Papanikolaou et al. (2011) |
| Waste cooking olive oil | *A. niger* NRRL363 | Filtration | SmF-250 mL flasks | 28°C, pH 6, 200 rpm | 8 | 1.15 | 0.64 | 0.74 | NR | NR | Papanikolaou et al. (2011) |
| FW | *Schizochytrium mangrovei* | Fungal hydrolysis by *A. oryzae* and *A. awamori*, autolysis | SmF-2 L bioreactor | 25°C, pH 6.5, 400 rpm | 4 | NR | 0.321 | NR | NR | 0.196 | Pleissner et al. (2013) |
| FW | *Chlorella pyrenoidosa* | Fungal hydrolysis by *A. oryzae* and *A. awamori*, autolysis | SmF-2 L bioreactor | 28°C, pH 6.5, 400 rpm | 4 | NR | 0.208 | NR | NR | 0.046 | Pleissner et al. (2013) |

*Source:* Reprinted from *Fuel*, 134, Uçkun, K. E. et al., Bioconversion of food waste to energy: A review, 389–399, Copyright 2014c, with permission from Elsevier.

*Note:* FW: food waste, Y: yield, P: productivity, SmF: submerged fermentation, μ: specific growth rate, A.: *Aspergillus*.

associated with the collection/transportation of FW should also be taken into account. Nevertheless, the low or no cost of food waste along with the environmental benefits considering the waste disposal would balance the initial high capital costs of the biorefineries. The efficiency and cost base of the production could be further improved by intensifying research and optimization studies on integrating different value-added product processes.

# 2 Platform Chemical Production from Food Wastes

## 2.1 INTRODUCTION

The amount of FW is increasing due to the rapid growth of global population and economics, particularly in Asian countries. The annual amount of urban FW generated in Asian countries will rise from 278 to 416 million tons from 2005 to 2025 (Melikoglu et al., 2013b). The highest absolute amount per year was in China (82.8 million tons [MT]) followed by Indonesia (30.9 MT), Japan (16.4 MT), the Philippines (12 MT), and Vietnam (11.5 MT). However, the highest amount of FW produced per capita was in New Zealand and Australia with 280 kg/year, while it was around 120–130 kg in Southeast Asia, except in Cambodia (173 kg/year). Although the absolute amount of FW in China is the highest, the waste production per capita is the lowest (61 kg/year), while the waste production per capita is 120 and 168 kg/year in Singapore and Hong Kong, respectively (National Environment Agency [Singapore], 2013; Lin et al., 2013), showing that FW seems more prevalent in high-income states. FW are dumped, landfilled, incinerated, composted, digested anaerobically, and/or used as animal feed. In many Asian countries, FW is still dumped with other household waste in landfills or dumpsites. Unfortunately, the capacity of landfills is limited, and increasing the amount of FW is not sustainable (Ngoc and Schnitzer, 2009).

In order to reduce its volume, FW is traditionally incinerated with other combustible municipal solid wastes for generation of heat or energy, particularly in Japan and Singapore. The incineration of FW is a favorable option against landfilling in terms of overall energy recovery, and emissions of greenhouse gases (Othman et al., 2013). However, the incineration may not be an affordable approach for most low-income countries due to the high capital and operating costs (Ngoc and Schnitzer, 2009). It should also be noted that incineration of FW can potentially cause air pollution (Takata et al., 2012). Another approach to handle biodegradable FW is composting, which results in a valuable soil conditioner and fertilizer (Gajalakshmi and Abbasi, 2008). Composting has a relatively low environmental impact and a high economic efficiency compared to other treatment methods. However, the high moisture content of FW may lead to substantial release of leachate (Cekmecelioglu et al., 2005). Compost is more expensive than commercial fertilizers and the current market of compost produced from FW is limited (Aye and Widjaya, 2006).

Anaerobic digestion is another alternative that yields methane and carbon dioxide as metabolic end products, and is therefore feasible from an economic and environmental point of view because methane is used as an energy source (Othman et al., 2013). Hirai et al. (2001) evaluated the environmental impacts of FW treatment and found that utilizing a methane fermentation process prior to incineration reduces approximately 70 kg $CO_2$ eq/ton waste of the global warming potential, due to the substitution effect. Food waste is also used as animal feed. The disadvantages are its variable composition and the high moisture content, which favors microbial contamination (Esteban et al., 2007). To prevent this, animal feed is generally dried but greenhouse gas emission increases depending on the energy usage during the drying process, which is related to the water content of FW (Takata et al., 2012). On the other hand, it should be noted that FW can be a potential resource for production of high-value chemicals (Lin et al., 2013). In general, organic acids, biodegradable plastics, and enzymes applications (selling price: $1000/ton biomass) usually create more value compared to generating electricity ($60–150/ton biomass) and animal feed ($70–200/ton biomass) and fuel applications ($200–400/ton biomass) (Sanders et al., 2007).

FW is mainly composed of carbohydrate (starch, cellulose, and hemicelluloses), proteins, lipids, and organic acids. Total sugar and protein contents in FW are in the range of 35.5%–69% and 3.9%–21.9%, respectively (Uçkun et al., 2014c). The hydrolysis of FW can be achieved using acids or alkali at ambient or high temperatures, leading to the breakage of glycosidic bonds with polysaccharides emerging as oligosaccharides and/or maltodextrins and monosaccharides. However, some substances inhibitory to subsequent fermentation may also be released. Since FW contains little cellulose, hemicelluloses, and lignin, enzymatic fermentation is the preferred route for its hydrolysis. Use of commercial enzymes for hydrolysis of FW has the drawbacks of high cost and less efficiency due to the fact that no tailored mixed commercial enzymes are available. In addition, each commercial enzyme requires different operating conditions for the hydrolysis of their specific substrates. Therefore, the process would either operate suboptimally with a mixture or take a long time to carry out each enzymatic step sequentially. Furthermore, FW composition varies constantly, which would make the mixture of commercial enzymes inefficient.

Alternatively, enzymes may be produced *in situ* by a fungus that can hydrolyze starch, a main component of FW. The resulting simple sugars are more amenable to fermentation into valuable platform chemicals, such as lactic acid, citric acid, succinic acid, and biopolymers such as polyhydroxyalkanoates (PHA) and polyhydroxybutyrates (PHBs). Compared to other agro-industrial raw materials, such as wheat bran, sugarcane bagasse, corn cob, and rice bran, FW is a better raw material for microbial fermentation without additional nutrient supplements. As such, FW has been used as the sole microbial feedstock for the development of various value-added bioproducts, for example, biofuels, biodiesel, platform chemicals, and enzymes (Han and Shin, 2004b; Sakai and Ezaki, 2006; Yang et al., 2006; Pan et al., 2008; Zhang et al., 2010b, 2013b; He et al., 2012a,b; Vavouraki et al., 2014).

Therefore, this chapter focuses mainly on FW fermentation technologies developed around the world for production of various platform chemicals, such as lactic, citric, and succinic acids and biopolymers from food waste.

## 2.2 PLATFORM CHEMICAL PRODUCTION

Platform chemicals are the main feedstocks for producing secondary chemicals, chemical intermediates, and final products (Jang et al., 2012). For example, succinic and lactic acids are used as starting materials for the production of pyrolidones and bioplastics, respectively. Although most of the platform chemicals are produced from petroleum, biological production of platform chemicals from renewable feedstock is gaining interest due to the limited oil reserves, increasing prices, and environmental concerns (Lin et al., 2013; Koutinas et al., 2014). Biological production of platform chemicals from agricultural products, lignocellulosic materials, and waste biomass has been reported in the literature (Koutinas et al., 2007c; Smith et al., 2010; Pfaltzgraff et al., 2013). Food waste is posing a global challenge due to its large volume and unstable nature. Therefore, the valorization of food waste is gaining more and more interest. The reported studies related to the production of platform chemicals from food waste are discussed in detail in the following sections.

### 2.2.1 LACTIC ACID

Lactic acid (LA) has been used for a long time in the food, pharmaceutical, textile, leather, and chemical industries (Yadav et al., 2011). Because LA has both carboxylic and hydroxyl groups, it can be converted into different potentially useful chemicals such as pyruvic acid, acrylic acid, 1,2-propanediol, and lactate ester (Fan et al., 2009). LA has two enantiomers, L (+) and D (−). Only L(+) isomer can be used in the food industry since the D(−) isomer is harmful to humans (Expert Committee on Food Additives, 1967). Nowadays, L-LA production is gaining interest due to its usage as a starting material to produce biodegradable plastics (poly-lactic acid, PLA) to be used as an alternative to synthetic plastics owing to its biodegradable and biocompatible nature (Chen and Patel, 2012). Each year 259,000 metric tons of LA is produced with an increasing trend, while the market price is about US$1300–1600 per ton (Wee et al., 2006).

In the LA production process, carbohydrates such as glucose and sucrose are metabolized by lactic acid bacteria. In addition, various nutrients such as yeast extract, peptone, or corn steep liquor are added as nitrogen sources as lactic acid bacteria have limited ability to synthesize amino acids and vitamin B. Many cheap, renewable raw materials such as molasses, starch, and lignocellulosic wastes from agricultural and agro-industrial residue have been used as substrates for LA fermentation (Gullon et al., 2008; Wang et al., 2010b). However, most starchy and lignocellulosic materials must be pretreated by physicochemical and enzymatic methods because LA-fermenting microorganisms cannot directly use these materials (Gao et al., 2011). In addition, most of the fermentative studies using waste materials have required additional nitrogen-rich substrates. From an economical viewpoint, it is more advantageous to produce LA from FW as it contains sufficient carbohydrates and other nutrients for microbial proliferation (Wang et al., 2005b; Sakai and Yamanami, 2006; Zhang et al., 2008; Wang, 2011). Moreover, organic acids such as LA followed by acetic and propionic acids were found to be the main products of kitchen waste fermentation (Wang et al., 2002). Therefore, it is possible to produce high yields of LA from FW (Sakai et al., 2004b; Ohkouchi and Inoue, 2006, 2007; Sakai and Ezaki, 2006; Omar et al., 2011). The studies conducted using FW and the LA yields achieved are summarized in Table 2.1. The highest LA yield from FW (1.29 g/g sugar) was reported by Sakai and Yamanami (2006). They obtained 40 g/L LA with 97% optical activity and 2.5 g/Lh productivity using a thermotolerant LA bacterium, *Bacillus licheniformis* TY7.

FW sterilization and inoculum addition are the important parameters, but they may increase the process cost of LA production. Lactic acid bacteria have an optimal fermentation temperature of 30–42°C. Therefore, it is difficult to avoid contamination if the medium is not sterilized. However, most of the studies were conducted using open fermentation (non-sterile) to decrease the process cost, improve the hydrolysis, and prevent the undesired chemical changes in FW due to sterilization (Sakai and Ezaki, 2006; Zhang et al., 2008; Wang et al., 2011). Sakai and Ezaki (2006) reported that LA selectively accumulated in the cultures of non-autoclaved FW using open fermentation by intermittent pH neutralization. However, the optical activity of accumulated LA was low because of the selective proliferation of *Lactobacillus plantarum*, a D- and L-LA producer (Sakai et al., 2004a). The isomer purity also depends on the medium pH and temperature (Akao et al., 2007; Zhang et al., 2008). Zhang et al. (2008) reported that the isomer purity was much higher at acidic or alkaline pH (non-controlled pH, pH 5, and pH 8) than neutral pH (pH 6 and pH 7). Increasing the fermentation temperature from 35°C to 45°C at pH 7 enhanced the isomer purity from 60:40 to 83:17 due to the change in the composition of microbial community. The LA bacteria and *Clostridium* sp. dominated the fermentation of non-sterile FW. The emergence and disappearance of LA bacteria resulted in the variations of the isomer purity. According to Akao et al. (2007), *Bacillus coagulans*, *Lactobacillus amylolyticus*, and *Clostridium thermopalmarium* produced L-lactate, racemic lactate, and butyrate, respectively, from non-sterilized FW, while their growth ceased by setting a cultivation temperature of 55°C and a pH below 5.5.

The effects of pH, temperature, oxygen, and the addition of minerals and nutrients on LA yield from FW has been reported (Idris and Suzana, 2006; Ohkouchi and Inoue, 2006, 2007; Ye et al., 2008b; Hafid et al., 2011; Wang et al., 2011). According to Hafid et al. (2011) the most significant parameters affecting the bioconversion of FW to organic acids were temperature and inoculum size. The highest level of organic acids (77 g/L with 78% LA) was produced at optimum pH, temperature, inoculum size of 6.02, 35–37°C, and 20% inoculum, respectively. Ohkouchi and Inoue (2006) reported that the productivity was affected by initial pH, culture pH, and manganese concentration. A yield of 1.11 g LA/g FW was obtained during the fermentation at pH 5. The addition of manganese improved the growth of *Lactobacillus* sp. and resulted in complete bioconversion within 100 h. Ohkouchi and Inoue (2007) have also determined that the bioconversion of sugars to LA was affected by the ratio of total sugars to total nitrogen content (TS/TN ratio). LA production was improved by nitrogen supplementation to adjust the TS/TN ratio below 10. In another study, Idris and Suzana (2006) investigated the effects of sodium alginate concentration, bead diameter, initial pH, and temperature on LA production from pineapple waste using immobilized *Lactobacillus delbrueckii*. The maximum LA yield (0.82 g LA/g FW) was obtained using 2% sodium alginate with a 1.0 mm bead diameter at an initial

**TABLE 2.1**

**Lactic Acid Production from Food Wastes**

| Waste | Microorganism | Pretreatment | Vessel Type | Conditions | Duration (days) | Y (g/g waste) | Y (g/g sugar) | P (g/Lh) | Selectivity (%) LA | References |
|---|---|---|---|---|---|---|---|---|---|---|
| KW | From unsterilized KW | Minced, autoclaved | 50 mL centrifuge tube with 30 mL working vol. | 37°C, pH 7 static | 5 | 0.1 | NR | 1.05 | 91 | Sakai et al. (2000) |
| Shrimp waste | Lactobacillus spp. Strain B2 | Minced | 30 kg solid state column reactor | 30°C, initial pH 7.5 | 6 | NR | NR | NR | NR | Cira et al. (2002) |
| FW | L. delbrueckii NRRLB-445 | Minced autoclaved | 250 mL flask with 100 mL working vol. | 42°C, initial pH 6, 150 rpm | 5 | 0.45 | NR | NR | NR | Kim et al. (2003) |
| FW | L. rhamnosus KY-3 | Propionic acid fermentation. glucoamylase pretreatment | 90 L fermenter | 37°C, pH 6.5 | 4 | 0.55 | 0.84 | NR | NR | Sakai et al. (2004b) |
| KW | L. sp. TH165 and 175 | Minced | 500 mL reactor with 250 mL working vol. | 37°C, pH 5.5–6 | 3 | 0.46 | 0.73 | 0.47 | NR | Wang et al. (2005a) |
| KW | L. plantarum TD46 | Minced | 500 mL reactor with 250 mL working vol. | 30°C, pH 5.5–6 | 2 | 0.39 | 0.62 | 0.6 | NR | Wang et al. (2005b) |
| Pineapple waste | L. delbrueckii ATCC9646 | Ground, filtered | 250 mL flask with 100 mL working vol. | 37°C, pH 6.5, 150 rpm | 3 | 0.48 | 0.82 | NR | NR | Idris and Suzana (2006) |
| FW | L. salivarus | Ground, heat treated at 80°C for 30 min. | 2 kg capacity plastic containers | 25°C, pH 5.98 | 30 | NR | NR | NR | NR | Yang et al. (2006) |
| FW | L. manihotivorans LMG18011 | Ground, autoclaved | NR | 30°C, pH 5–5.5 | 4 | NR | 1.11 | NR | NR | Ohkouchi and Inoue (2006) |
| Shrimp waste | L. plantarum 541 | Crushed | 1 L beaker with 200 g working vol. | 30°C, pH 6 | 2 | NR | NR | NR | NR | Rao and Stevens (2006) |

(Continued)

**TABLE 2.1** (*Continued*)
**Lactic Acid Production from Food Wastes**

| Waste | Microorganism | Pretreatment | Vessel Type | Conditions | Duration (days) | Y (g/g waste) | Y (g/g sugar) | P (g/Lh) | Selectivity (%) LA | References |
|---|---|---|---|---|---|---|---|---|---|---|
| Model KW | *Bacillus cuagulans* NBR12583 | Minced, glucoamylase pretreated, filtered | 2 L fermenter with 1.2 L working vol. | 50°C, pH 6.5, 60 rpm | 5 | NR | 0.98 | NR | NR | Sakai and Ezaki (2006) |
| Model KW | *L. plantarum* | Minced | 50 mL tubes with 30 mL working vol. | 37°C, pH 6.2 | 5 | NR | NR | NR | 95 | Sakai et al. (2006) |
| Model KW | Thermotolerant *Bacillus licheniformis* TY7 | Minced, glucoamylase pretreated, filtered | 2 L fermenter with 1.2 L working vol. | 50°C, pH 6.2 | 3 | NR | 1.29 | 2.5 | 100 | Sakai and Yamanami (2006) |
| Synthetic FW | None | Milled | 1 L working volume | 55°C, pH 5.5 | 12.5 | NR | 0.5 | NR | NR | Akao et al. (2007) |
| Apple pomace | *L. rhamnosus* ATCC9595 | Dried, milled, autoclaved, cellulase and glucosidase hydrolysis | Bioreactor | 37°C, pH 5–6, 150 rpm | 4 | 0.366 | NR | 2.78 | NR | Gullon et al. (2007) |
| KW | *L. manihotivorans* LMG18011 | Ground, autoclaved | Bioreactor with 400 mL working vol. | 25°C, pH 5 | 6 | 0.13 | 1.08 | NR | NR | Ohkouchi and Inoue (2007) |
| Apple pomace | *L. rhamnosus* ATCC9595 | Dried, milled, autoclaved, cellulase and glucosidase hydrolysis | Bioreactor | 41.5°C, pH 5.8, 150 rpm | 4 | 0.465 | 0.88 | 5.41 | NR | Gullon et al. (2007) |
| FW | *L. rhamnosus* 6003 | Milled, autoclaved, glucoamylase pretreated | 500 mL erlenmeyer flask with 250 mL working vol. | 35°C, 120 rpm | 2.5 | 0.455 | NR | NR | NR | Wang et al. (2009b) |

(Continued)

**TABLE 2.1 (Continued)**
**Lactic Acid Production from Food Wastes**

| Waste | Microorganism | Pretreatment | Vessel Type | Conditions | Duration (days) | Y (g/g waste) | Y (g/g sugar) | P (g/Lh) | Selectivity (%) LA | References |
|---|---|---|---|---|---|---|---|---|---|---|
| FW | L. plantarum BP04 | Dried, milled, amylase and protease pretreated | 250 mL flask with 100 mL working vol. | 35°C, pH 6, 120 rpm | 1 | NR | NR | NR | NR | Ye et al. (2008a) |
| FW | None | Ground | 100 mL bottle with 80 mL working vol. | 45°C, pH 7 | 6 | NR | NR | NR | NR | Zhang et al. (2008) |
| FW | None | Ground | 50 L stirred tank reactor with 30 L working vol. | 37°C, initial pH 6, 150 rpm | 7 | 1.153 | NR | NR | 98 | Omar et al. (2009) |
| KW | Naturally fermented KW | Ground | 250 mL flask with 100 mL working vol | 37°C, pH 5, 200 rpm | 10 | NR | 0.43 | 0.31 | 76.2 | Hafid et al. (2010) |
| KW | Naturally fermented KW | Ground | 250 mL flask with 100 mL working vol. | 35°C, pH 6.02, 200 rpm | 5 | NR | NR | NR | 78 | Hafid et al. (2010) |
| KW | Lactobacillus TY50 | Ground | 500 mL erlenmeyer flask with 250 mL working vol. | 45°C pH 5.5–6, 100 rpm | 2 | 0.44 | 0.7 | 1.01 | NR | Wang et al. (2011) |
| FW and sludge | None | Ground | Reactor with 1.2 L working vol. | 21°C, pH 8, 120 rpm | 8 | NR | NR | NR | NR | Chen et al. (2013) |
| Mango peel waste | None | Ground, autoclaved | 100 mL bottle with 80 mL working vol. | 35°C, pH 10 | 6 | NR | NR | NR | NR | Jawad et al. (2013) |

Source: Uçkun, E. K., A. P. Trzcinski, and Y. Liu: Platform chemicals production from food wastes using a biorefinery concept. *Journal of Chemical Technology and Biotechnology.* 2015. 90 (8). 1364–1379. Copyright Wiley-VCH Verlag GmbH & Co. KGaA. Reproduced with permission.

Note: L.: Lactobacillus, FW: food waste, KW: kitchen waste, LA: lactic acid, Y: yield, P: productivity.

pH of 6.5 and temperature of 37°C. Wang (2011) investigated the effects of temperature, pH, and oxygen on LA production from FW in an open fermentation using *Lactobacillus* TY50 as inoculum. The optimum temperature and pH were 45°C and pH 6 for a synergistic relationship between the inoculated strain and indigenous strains, resulting in higher LA yield (0.44 g LA/g FW) and productivity (1.01 g/L · h).

In other studies, the conversion of FW into LA was improved via simultaneous saccharification and fermentation using different enzymes/enzyme mixtures or enzyme-producing microorganisms with LA bacteria. Kim et al. (2003) investigated the effect of simultaneous saccharification of the starch component in FW by a commercial amylolytic enzyme preparation (a mixture of amyloglucosidase, α-amylase, and protease) and fermentation by *Lactobacillus delbrueckii*. The highest LA yield was 0.45 g LA/g FW using 42°C and pH 6, without supplementation of nitrogen-containing nutrients and minerals. LA biosynthesis was also reported to be significantly affected by protease pretreatment and temperature while α-amylase treatment had no significant effects (Ye et al., 2008b).

In a study by Wang et al. (2008b), instead of utilizing commercial saccharification enzymes, *Aspergillus niger* was grown in solid state fermentation (SSF) to produce glucoamylase. The crude glucoamylase can be used directly for the saccharification of FW prior to LA fermentation. LA yield (0.455 g/g FW) was improved by 72% compared to the results obtained without glucoamylase indicating that it may be more economically efficient to integrate *in situ* enzymes production and fermentation processes. In another study, Sakai et al. (2004b) developed an integrated novel system for the conversion of municipal FW into poly-L-lactate (PLLA) biodegradable plastics. The process consists in the removal of endogenous LA from FW by *Propionibacteria*, LA fermentation under semisolid conditions, L-LA purification via butyl esterification, and L-LA polymerization. High yield of L-LA and PLLA (0.55 and 0.32 g/g dry FW, respectively) with high optical activity (i.e., a high proportion of optical isomers) were obtained. This design also enabled recycling of all materials produced at each step with energy savings and minimal emissions. The physical properties of PLLA were comparable to those of PLLA generated from commercially available L-LA, indicating that FW is a good candidate for biodegradable plastic production.

## 2.2.2 CITRIC ACID

Citric acid (CA), a natural metabolic intermediate of Kreb's cycle, is generally recognized as safe, nontoxic, and biodegradable. It is used extensively in the food, pharmaceutical, cosmetic, agricultural, and biochemical industries (Kamzolova et al., 2008; Gurpreet et al., 2011; Hamdy, 2013). It is also used in many healthcare products because of its antioxidant properties. Nowadays, CA is gaining popularity in the biomedicine sector, where it is used as an active ingredient in the formation of different biopolymers for use in nano-medicines, drug delivery systems, and for culturing a variety of human cells (Dhillon et al., 2011). Therefore, its consumption is escalating gradually and the global production reached 1.4 million tons per year (Hamdy, 2013). Microbial production of CA is a well-established method for obtaining this important commodity chemical from renewable resources. A variety of substrates, such as sucrose, starch from various sources, cane and beet molasses, fruit wastes such as apple pomace, date waste, orange/citrus wastes, and banana peel have been utilized for the economical production of CA (Dhillon et al., 2011; Angumeenal and Venkappayya, 2013). To date, there is no publication on CA production from kitchen refuse or mixed FW. The yields achieved in recent CA production studies are summarized in Table 2.2. Most of the studies were conducted using *Aspergillus niger* and *Yarrowia lipolytica* sp. at 30°C, pH 3–6.5. Among the studies, the highest CA yield (0.8 g CA/g sugar) was reported by Shojaosadati and Babaeipour (2002). They produced CA with a yield of 80% based on total sugars from apple pomace using *Aspergillus niger* in a packed bed bioreactor. The optimum conditions were obtained using an aeration rate of 0.8 L.min$^{-1}$, particle size of 0.6–2.33 mm, and a moisture content of 78% (w/w).

Methanol is widely used as an inducer in the microbial production of CA due to its stimulatory effects (Kumar et al., 2003, 2010; Alben and Erkmen, 2004; Rivas et al., 2008; Dhillon et al., 2011; Hamdy, 2013). A methanol concentration of 3%–4% is commonly used. Rodrigues et al. (2009) demonstrated that methanol addition decreased the inhibition caused by heavy metals present in the fermentation medium. It also affects the permeability properties of the mold and enables greater excretion of CA through the membrane. Moreover, methanol also acts as an inducer for the enzyme citrate synthase, which ultimately leads to greater CA production (Barrington and Kim, 2008). Dhillon et al. (2011) achieved 0.342 g/g substrate yield in SSF of apple pomace using *Aspergillus niger* by supplementing 3% methanol.

In order to improve the efficiency of the CA production process, some strategies such as co-utilization of different wastes, inoculation of mutant strains, co-culture of microorganisms/strains, and/or integration to other processes were studied. Hamdy (2013) improved the CA production by fortifying orange peel medium with an industrial by-product, cane molasses. As a result, CA yield improved by 40%.

Karasu Yalcin (2012) cultured *Yarrowia lipolytica* strains by mutations via UV-irradiation and/or ethyl methane sulfonate, and further used the selected strains to produce CA from celery and carrot wastes. Chemical mutagenesis was found to be more effective in enhancing CA production than UV-induced mutagenesis. Maximum CA concentration (50.1 g/L) and yield obtained by the chemical mutant *Y. lipolytica* K-168 exceeded that of the initial strain by 57%. In another study, the efficiency of CA production was improved by the integrated cellulase and pectinase production from potato wastes using *Saccharomyces cerevisiae* (Afifi, 2011). The pectinase and cellulase pretreated potato wastes resulted in CA yields of 0.47 and 0.68 g/g, respectively, using *A. niger* MAF3. FW therefore offers an opportunity to produce high activity cellulase and pectinase, as well as CA to be used in various industries.

### 2.2.3 Succinic Acid

Succinic acid (SA) is one of the top commodity chemicals that forms the basis for supplying many important intermediate and specialty chemicals for the consumer products industry (Bozell and Petersen, 2010). As a commodity chemical, SA could replace many commodities based on benzene and intermediate petrochemicals, resulting in

**TABLE 2.2**

**Citric Acid Production from Food Wastes**

| Waste | Microorganism | Pretreatment | Vessel Type | Conditions | Duration (days) | Y (g/g waste) | Y (g/g sugar) | P (g/Lh) | References |
|---|---|---|---|---|---|---|---|---|---|
| Apple pomace | A. niger BC1 | Dried, ground | Packed bed reactor | 30°C, 78% MC (w/w), 0.6–2.33 mm PS, 0.8 L/min AR, SSF | 5 | 0.124 | 0.8 | NR | Shojaosadati and Babaeipour (2002) |
| Fruit waste | A. niger DS1 | Dried, ground | 250 mL flask with 5 g waste | 30°C, 60% MC, 1.2–1.6 mm PS, 4% methanol, SSF | 8 | 0.107 | 0.465 | NR | Kumar et al. (2003) |
| Semolina waste | A. niger ATCC9142 | Autoclaved | 250 mL flask with 150 mL working vol. | 30°C pH 6.5%, 3% methanol, SmF | 14 | NR | 0.219 | NR | Alben and Erkmen (2004) |
| Orange peel | A. niger ATCC9142 | Dried, ground, autohydrolysis (130°C) | 100 mL flask with 40 g working vol. | 30°C, 200 rpm, SmF | 6 | NR | 0.53 | 0.128 | Rivas et al. (2008) |
| Pineapple waste | Yarrowia lipolytica NCIM3589 | Dried, ground | 250 mL flask with 20 mL working vol. | 30°C, pH 6.4–6.8%, 70% MC, SSF | 6 | 0.202 | NR | NR | Imandi et al. (2008) |
| Banana peel | A. niger MTCC282 | Steamed, homogenized | Glass trough having 30 cm diameter | 28°C, pH 3–5%, 70% MC, SSF | 3 | 0.18 | NR | NR | Karthikeyan and Sivakumar (2010) |
| Apple pomace | A. niger van. Tieghem MTCC 281 | Ground | 1 L flask with 100 g working vol. | 30°C, 4% methanol, SmF | 5 | 0.05 | NR | NR | Kumar et al. (2010) |
| Orange peel | A. niger ATCC9142 | Dried, ground | 50 mL flask with 1 g working vol. | 30°C, 70% MC, SSF | 5 | 0.193 | 0.61 | NR | Torrado et al. (2011) |

*(Continued)*

**TABLE 2.2 (Continued)**
**Citric Acid Production from Food Wastes**

| Waste | Microorganism | Pretreatment | Vessel Type | Conditions | Duration (days) | Y (g/g waste) | Y (g/g sugar) | P (g/Lh) | References |
|---|---|---|---|---|---|---|---|---|---|
| Apple pomace | A. niger NRRL-567 | Dried, ground | 12 L tray fermenter with 1 kg working vol. | 30°C, pH 3.43%, 3% methanol, SSF | 7 | 0.342 | NR | NR | Dhillon et al. (2011) |
| Potato waste | A. niger MAF3 | Co-cultured with S. Cerevisiae | 250 mL flask with 50 mL working vol. | 30°C, SmF | 5 | 0.068 | NR | NR | Afifi (2011) |
| Date waste | A. niger ATCC16404 | Heated, filtered, decanted | 3 L reactor with 1 L working vol. | 30°C, pH 3.5, 200 rpm, 1 L/L min AR, 3% methanol, SmF | 6 | NR | 0.76 | NR | Acourene et al. (2011) |
| Date waste | A. niger ANSS-B5 | Heated, filtered, decanted | 3 L reactor with 1 L working vol. | 30°C, pH 3.5, 200 rpm, 3% methanol, SmF | 5 | NR | 0.67 | NR | Acourene and Ammouche (2012) |
| Carrot and celery by-products | Yarrowia lipolytica 57 | Smashed, filtered | 300 mL flask with 100 mL working vol. | 30°C, pH 5.5, 100 rpm, SmF | 10 | NR | NR | NR | Karasu Yalcin (2012) |
| Orange peel | A. niger van Tiegh 1867 | Dried, ground | 250 mL flask with 50 mL working vol. | 30°C, pH 5, 250 rpm, 65% MC, 3.5% methanol, SmF | 3 | 0.38 | NR | NR | Hamdy (2013) |

*Source:* Uçkun, E. K., A. P. Trzcinski, and Y. Liu: Platform chemicals production from food wastes using a biorefinery concept. *Journal of Chemical Technology and Biotechnology.* 2015. 90 (8). 1364–1379. Copyright Wiley-VCH Ver ag GmbH & Co. KGaA. Reproduced with permission.

*Note:* AR: aeration rate, PS: particle size, MC: moisture content, CA: citric acid, A.: *Aspergillus*, SSF: solid state fermentation, SmF: submerged fermentation, Y: yield, P: productivity, NR: Not reported.

a large reduction in the consumption of over 250 benzene-derived chemicals (Zeikus et al., 1999). Chemicals produced from SA include γ-butyrolactone, tetrahydrofuran, 1,4-butanediol, adipic acid, succinonitrile, succindiamide, 4,4-Bionolle®, and various pyrrolidones (Lin et al., 2013). The annual production of SA reached 30,000–35,000 tons with a market value of $225 million and is continuing to increase gradually (Taylor, 2010). Therefore, there is a growing research interest in the development of fermentative SA production as an alternative to the current non-sustainable petrochemical route. The most commonly used microorganisms for SA production are *Actinobacillus succinogenes*, *Anaerobiospirillum succiniciproducens*, *Mannheimia succiniciproducens*, and recombinant *Escherichia coli* strains (Zeikus et al., 1999; Bozell and Petersen, 2010). Theoretically, two molecules of SA are produced from one molecule of monosaccharide in fermentation (Zeikus et al., 1999). The main advantage of SA fermentation over other types of fermentation is that it can be produced anaerobically and will consume carbon dioxide (one mole $CO_2$ per mole SA) making it environmentally friendly. The possibility to convert FW into SA has been explored. The recent publications and the yields achieved are summarized in Table 2.3. SA production was conducted on starch-rich FW such as wheat bran, bakery wastes such as pastry, bread, and cakes, and also from mixed food waste from restaurants.

Dorado et al. (2009) developed a wheat-based bioprocess for the production of a fermentative feedstock for SA production. Wheat bran was used as the sole medium in two SSF of *Aspergillus awamori* and *Aspergillus oryzae* that produce enzyme complexes rich in amylolytic and proteolytic enzymes, respectively. The resulting fermentation solids were then used as crude enzyme sources, to hydrolyze milled bran and middling fractions to generate a hydrolysate rich in glucose, maltose, and free amino nitrogen. This hydrolysate was then used for *Actinobacillus succinogenes* fermentations, which led to the production of 1.02 g SA/g sugar with a productivity of 0.91 g/Lh. Leung et al. (2012) used bread waste as a generic feedstock for the production of a nutrient rich biomedium. The fermentation of the bread hydrolysate by *Actinobacillus succinogenes* resulted in a yield and productivity of 1.16 g SA/g glucose and 1.12 g/Lh, respectively. This corresponds to an overall yield of 0.55 g/g bread. This is the highest SA yield compared from other FW-derived media reported to date. Assuming a SA yield from bakery waste of 1.16 g/g glucose and based on 1 ton/day of bakery waste, the potential SA production could be 25,000 kg per year for a city like Hong Kong with 7 million inhabitants (Lam et al., 2014). Zhang et al. (2013a) have also used the same strategy, and evaluated the potential of bakery waste, including cakes and pastries for SA production. Both hydrolysates were found to be rich in glucose and free amino nitrogen and therefore suitable as feedstock in *Actinobacillus succinogenes* fermentation.

The overall SA produced from pastry and cake wastes was 0.35 and 0.28 g/g substrate, respectively, and high recovery yields of SA crystals (96%–98%) were obtained. Mixed food waste directly collected from restaurants has been used for SA fermentation with the recombinant *E. coli* and *Actinobacillus succinogenes* at an overall yield of 0.224 g/g substrate (Lam et al., 2014). This implies that food waste is a potential renewable resource for SA production.

## TABLE 2.3
## Succinic Acid Production from Food Wastes

| Waste | Microorganism | Pretreatment | Vessel Type | Conditions | Duration (days) | Y (g/ g waste) | Y (g/ g sugar) | P (g/Lh) | References |
|---|---|---|---|---|---|---|---|---|---|
| Wheat bran | Actinobacillus succinogenes ATCC 55618 | SSF using A. awamori and A. oryzae | 1.8 L bioreactor with 0.8 L working vol. | 37°C, pH 6.6–6.8, 200 rpm, 30 g/L MgCO₃ | 2 | NR | 0.8 | 1.19 | Du et al. (2008) |
| Wheat bran | Actinobacillus succinogenes ATCC 55618 | SSF using A. awamori and A. oryzae | 1.8 L bioreactor with 0.8 L working vol. | 37°C, pH 6.6–6.8, 200 rpm, 10 g/L MgCO₃ | 3 | NR | 1.02 | 0.91 | Dorado et al. (2009) |
| Orange peel | Fibrobacter succinogenes S85 (ATCC 19169) | Ground, dried, limonene separation | 125 mL bottles anaerobic | 39°C, 25 rpm | 3 | NR | 0.12 | 0.025 | Li et al. (2010) |
| Bread waste | A. succinogenes ATCC 55618 | SSF using A. awamori and A. oryzae | 2.5 L bioreactor | 37°C, pH 6.6–6.8, 300 rpm, 10 g/L MgCO₃ | 2 | 0.55 | 1.16 | 1.12 | Leung et al. (2012) |
| Cakes | Actinobacillus succinogenes (ATCC 55618) | SSF using A. awamori and A. oryzae | 2.5 L bioreactor | 37°C, pH 6.6–6.8, 300 rpm, 10 g/L MgCO₃ | 2 | 0.28 | 0.8 | 0.79 | Zhang et al. (2013a) |
| Pastries | Actinobacillus succinogenes (ATCC 55618) | SSF using A. awamori and A. oryzae | 2.5 L bioreactor | 37°C, pH 6.6–6.8, 300 rpm, 10 g/L MgCO₃ | 2 | 0.35 | 0.67 | 0.87 | Zhang et al. (2013a) |

*Source:* Uçkun, E. K., A. P. Trzcinski, and Y. Liu: Platform chemicals production from food wastes using a biorefinery concept. *Journal of Chemical Technology and Biotechnology.* 2015. 90 (8). 1364–1379. Copyright Wiley-VCH Verlag GmbH & Co. KGaA. Reproduced with permission.

*Note:* SSF: solid state fermentation, A.: *Aspergillus,* Y: yield, P: productivity.

During SA production, $MgCO_3$ is supplemented to improve the yield. It is the most satisfactory neutralization reagent used for pH control because it prevents cell flocculation and prolongs the stationary phase (Du et al., 2008; Dorado et al., 2009). In another study, waste orange peelings were used to produce SA by a consolidated bioprocess that combines cellulose hydrolysis and fermentation, using a cellulolytic bacterium, *Fibrobacter succinogenes* S85 (Li et al., 2010). Orange peelings contain D-limonene, which can inhibit bacterial growth. Therefore, it was pretreated by steam distillation to remove D-limonene. Pretreated orange peelings generated a SA yield of 0.12 g/g sugar and a productivity of 25 mg/Lh.

From Table 2.3, it can be seen that SA fermentation is traditionally conducted in batches using pure substrates. The initial glucose concentration is generally around 60–70 g/L in order to obtain a final SA concentration of 30–40 g/L. Lower succinate concentrations would make the recovery inefficient and too costly. Since FW can be used to generate up to 90–140 g/L sugars, it has the potential to be used as feedstock for SA, but there is a lack of studies on the fermentation of mixed FW to SA. Furthermore, pure starchy waste may not be available and sorting of FW is not an option. More studies should therefore be carried out on SA production from mixed FW. Further research should also focus on genetically modified strains that can metabolize high sugar concentrations.

## 2.2.4 POLYHYDROXYALKANOATES

Polyhydroxyalkanoates (PHAs) are hydroxyalkanoate polyesters of various chain lengths that can be synthesized by various microorganisms as intracellular energy- and carbon-storage materials under growth-limiting and carbon excess conditions (Steinbuchel and Fuchtenbusch, 1998). PHAs and their derivatives have numerous potential uses such as specialty biopolymer for medical applications and production of the platform molecule 3-hydroxybutyric acid (Steinbuchel and Fuchtenbusch, 1998; Novikova et al., 2008; Mooney, 2009). PHAs includes polyhydroxybutyric acid (PHB), polyhydroxyvaleric acid (PHV), 3-hydroxy-2-methylbutyrate (3H2MB), and 3-hydroxy-2-methylvalerate (3H2MV) (Rhu et al., 2003). PHAs have physical properties that range from brittle and thermally unstable to soft and tough depending on their butyrate PHB to PHV ratio (Zhang et al., 2013c). PHB has been proposed as an alternative to conventional petroleum-based plastics due to its thermoplasticity, 100% resistance to water, and 100% biodegradability (Sakai et al., 2004b; Solaiman et al., 2006). Its elasticity is improved by its co-polymerization with PHV (Castilho et al., 2009). However, due to their high production costs, 8–10 times more than that of conventional plastics, the industrial production of PHAs is hindered (Zhang et al., 2013c). Thus, with the aim of commercializing PHAs, a substantial effort has been devoted to reduce the production cost through the development of bacterial strains, more efficient fermentation/recovery processes, and cheap substrates (Xu et al., 2010; Eshtaya et al., 2013).

Table 2.4 lists the studies of PHA production from FW. PHAs are generally produced using the organic acids generated by acidogenesis of FW, either by acid pretreatment or by anaerobic digestion by the native/supplemented microorganisms. The theoretical PHB yield was reported as 0.44–0.48 g/g glucose (Yamane, 1993).

**TABLE 2.4**
**Polyhdroxyalkanoates Production from Food Wastes**

| Product | Waste | Microorganism | Pretreatment | Vessel Type | Conditions | Duration (hours) | Y (g/ g waste) | Y (g/ g sugar) | Y (g/ g cell) | P (g/Lh) | References |
|---|---|---|---|---|---|---|---|---|---|---|---|
| PHB | Soya waste | Alcaligenes latus DSM1124 | Acidic pretreatment, filtration, neutralization | 3.7 L fermenter with 2.4 L working vol. | 35°C, pH 7, DO: 20% | 51 | NR | NR | 0.11 | 0.11 | Yu et al. (1999) |
| PHB | Potato waste | Alcaligenes eutropHus | Enzymatic saccharification and liquefaction | Bioreactor with 1 L working vol. | 30°C, pH 7 | 54 | NR | 0.37 | 0.77 | NR | Rusendi and Sheppard (1995) |
| PHB | Soya waste | Rec. E. coli | Acidic pretreatment, filtration, neutralization | Bioreactor with 3 L working vol. | 37°C, pH 7, 200 rpm | 18 | NR | NR | 0.23 | NR | Hong et al. (2000) |
| PHA | FW | Ralstonia eutropha ATCC17699 | Acidogenesis without inoculum addition | 1.6 L air bubbling bioreactor with 1.3 L working vol. | 30°C, pH 7.5, DO:20% | 80 | NR | NR | 0.73 | NR | Du and Yu (2002) |
| PHA | FW | SS | Acidogenesis with SS | SBR with 4 L working vol. | 35°C | 60 | 0.025 | NR | 0.51 | NR | Rhu et al. (2003) |
| PHA | FW | Ralstonia eutropha ATCC17699 | Acidogenesis without inoculum addition | 2 L air bubbling bioreactor | 30°C, pH 7.5, DO:20% | 80 | NR | NR | 0.73 | NR | Du et al. (2004) |

*(Continued)*

**TABLE 2.4 (Continued)**
**Polyhydroxyalkanoates Production from Food Wastes**

| Product | Waste | Microorganism | Pretreatment | Vessel Type | Conditions | Duration (hours) | Y (g/ g waste) | Y (g/ g sugar) | Y (g/ g cell) | P (g/Lh) | References |
|---|---|---|---|---|---|---|---|---|---|---|---|
| PHB | Wheat milling by-products | *Wautersia eutropha* NCIMB 11599 | SSF using *A. Awamori* | 1.5 L bioreactor | 30°C, pH 6.7–6.9, 200 rpm, DO:40% | 62 | 0.3 | 0.47 | 0.93 | 0.9 | Xu et al. (2010) |
| PHB | FW | *Cupriavidus necator* H16 | Acidogenesis without inoculum addition | 1.5 L air bubbling bioreactor | 20°C, pH 7.5 | 259 | NR | NR | 0.87 | NR | Hafuka et al. (2011) |
| PHB | KW | *Cupriavidus necator* CCGUG 52238 | Acidogenesis without inoculum addition | 7 L bioreactor with 5 L working vol. | 30°C, pH 7, DO: 30% | 40 | NR | 0.38 | 0.86 | 0.24 | Omar et al. (2011) |
| PHB | FW | NR | Acidogenesis without inoculum addition | SBR bioreactor with 6 L working vol. | 30°C, 1000 rpm, DO:20% | 6 | NR | NR | 0.68 | NR | Md Din et al. (2012) |
| PHA | FW | NR | Ground, filtered, oil removal, acidogenesis with domestic sludge | 250 mL flask with 100 mL working vol. | 29°C, 100 rpm | 72 | NR | NR | 0.4 | NR | Venkateswar Reddy and Venkata Mohan (2012) |
| PHB | FW | Rec. *E. coli* | Acidogenesis | 500 mL flask | 37°C, pH 7, 200 rpm | 24 | NR | NR | 0.44 | 0.54 | Eshtaya et al. (2013) |
| PHA | FW+SS | SS | Crushed, mixed with SS | SBR bioreactor with 1 L working vol. | 25°C | 12 | NR | NR | 0.48 | NR | Zhang et al. (2013c) |

*Source:* Uçkun, E. K., A. P. Trzcinski, and Y. Liu: Platform chemicals production from food wastes using a biorefinery concept. *Journal of Chemical Technology and Biotechnology.* 2015. 90 (8). 1364–1379. Copyright Wiley-VCH Verlag GmbH & Co. KGaA. Reproduced with permission.

*Note:* FW: food waste, KW: kitchen waste, NR: not reported, DO: dissolved oxygen, SS: sewage sludge.

The highest yield (0.47 g/g glucose) was obtained by Xu et al. (2010) via fed-batch bioconversion of wheat milling by-products using *Wautersia eutropha*. Hafuka et al. (2011) reported a PHB yield of 0.38 g/g glucose (0.73 g/g dry biomass) in continuous fermentation by *Cupriavidus necator*. The PHB production ran for 259 hours by feeding the hydrolyzate through a 0.45-mm pore-size membrane filter without using any other expensive pretreatment. In a study by Du and Yu (2002), the mass transfer rates of fermentative acids and PHA production were enhanced using a dialysis membrane without washout of anaerobic microorganisms resulting in a PHB yield of 0.73 g/g dry biomass.

Studies on PHA synthesis from FW focused on the optimization of the operational conditions using different microbial cultures, feeding regimes, nutrient and oxygen supplementation, and recovery techniques to increase PHA yield. Most of the studies were conducted using pure cultures. Utilization of mixed cultures is more energy-efficient than pure cultures (Hafuka et al., 2011). However, the PHA content obtained from FW using mixed cultures was not high (<68%) while those using pure cultures yielded a very high PHA content (ca. 80%) (Table 2.4). Some recombinant microbial strains with improved PHB storage capabilities have also been used to obtain high yields of PHB in shorter time (Hong et al., 2000; Chien and Ho, 2008; Eshtaya et al., 2013). Eshtaya et al. (2013) reported the production of PHB from acidified FW (mainly lactic and acetic acids) with recombinant *E. coli*. A PHB yield of 0.44 g/g sugar and a productivity of 0.54 g/Lh were obtained within 24 hours.

Fermentation mode and feeding regimes can also affect PHB accumulation (Hafuka et al., 2011; Md Din et al., 2012). Omar et al. (2011) demonstrated that lactic acid concentration above 10 g/L inhibited both cell growth and PHB production in *Cupriavidus necator* CCGUG 52238. Hence, the highest PHB production was only 52.8% in batch fermentation with a yield and productivity of 0.38 g/g and 0.065 g/Lh, respectively. However, a PHB content of 84.5% and about fourfold increase in PHB productivity (0.24 g/Lh) was achieved by applying a fed-batch strategy. Hafuka et al. (2011) also investigated the effects of different feeding regimes on PHB production using *Cupriavidus necator*. The highest cell concentration was obtained using one-pulse feeding, while higher PHB concentration was achieved using stepwise- and continuous-feeding regimes. Therefore, the continuous-feeding regime was chosen for continuous PHB production.

It is also known that cells accumulate PHB under nitrogen- and/or oxygen-limited media with a sufficient carbon source (Jung and Lee, 2000; Rhu et al., 2003). Punrattanasin et al. (2006) reported that aerobic conditions combined with multiple nutrient limitations produced greater PHAs than strategies relying on oxygen limitation (either microaerophilic or anaerobic/aerobic). In another study, the microaerophilic conditions increased the PHA content in activated sludge to as much as 62% using acetate as the primary carbon source (Satoh et al., 1998). Venkateswar Reddy and Venkata Mohan (2012) also demonstrated that anoxic microenvironment provides higher PHA production, while aerobic microenvironment showed higher substrate degradation. FW composition, especially C/N ratio, also affects PHA accumulation. FW with low nitrogen content (C/N above 30) is suitable for PHA production (Omar et al., 2011). The acetogenesis step is another factor affecting the PHA composition. PHB proportions are higher when even numbers of carbon (acetate, butyrate) are used, whereas, 3HB-co-3HV was increased when odd numbers of carbon (propionate, valerate) were used (Akiyama et al., 1992; Rhu et al., 2003; Zhang et al., 2013c). Zhang et al. (2013c) reported that the composition of the liquid produced from FW by acidogenesis depends on the organic loading rate (OLR), FW composition, and pH while the latter is the most significant factor. By controlling the ratio of even-numbered to odd-numbered VFAs, the PHB content in the PHAs could be controlled within the range of 22%–30%.

Biodegradable polymers are promising materials and can be produced in large scale if the production yield is improved. Venkateswar Reddy and Venkata Mohan (2012) integrated the PHB production process to biohydrogen to improve their efficiencies. The main challenges encountered with fermentative $H_2$ are low substrate conversion efficiency and acid-rich residual effluents generated from the acidogenic process (Chu et al., 2008; Gómez et al., 2011). Hydrogen is considered an intermediate toward the production of VFAs, which is an important precursor for further conversion to PHAs. Integration with biohydrogen production yielded additional substrate degradation under both aerobic (78%) and anoxic (72%) microenvironments apart from PHA production.

## 2.2.5 SINGLE CELL OILS

Microbial lipids or single cell oils (SCOs) are the lipids produced by oleaginous microorganisms, such as yeasts, fungi, microalgae, and bacteria. These microorganisms can store more than 20% of their dry weight as lipids (Ratledge, 1991). SCOs can be used as a substituent of natural oils and fats for chemical production (Koutinas et al., 2014). The applications of SCOs depend on fatty acid compositions of SCOs (Kyle, 2001; Huang et al., 2013). SCOs are generally used for biodiesel production when the fatty acid composition is similar to vegetable oils (Uçkun et al., 2012). Alternatively, they are used for medical and dietetic purposes when SCOs contain polyunsaturated fatty acids, such as γ-linolenic acid, docosahexaenoic acid, eicosapentaenoic acid, and arachidonic acid (Ward and Singh, 2005), or used as cocoa-butter substitutes when the fatty acid composition is similar to that of cocoa-butter (Papanikolaou et al., 2003). The lipid metabolism and fatty acid composition of SCOs are influenced by the substrate used. To date, many low cost, hydrophilic (sugars, glycerol) and hydrophobic (fats, oils) substrates have been used to investigate economical SCO production. Lipid accumulation from sugars and/or glycerol usually requires nitrogen-limited culture conditions with excess carbon, known as *de novo* lipid accumulation process. When nitrogen is low, the key enzymes in the Krebs cycle cannot be activated. The accumulated citric acid is secreted in the cytoplasm and cleaved to acetyl-CoA and oxaloacetate, while the fatty acid synthetase catalyzes the conversion of acetyl-CoA to fatty acids, and then triglycerides are formed (Papanikolaou and Aggelis, 2011). On the contrary, *ex novo* lipids are produced through the fermentation of hydrophobic substrates such as fats and oils. The triglycerides found in the growth medium are converted to fatty acids by extracellular or cell-bound lipases, and then transported into the cell and used for growth (through β-oxidation), or accumulate as storage materials after/before the alteration by enzyme catalyzed reactions. The *ex novo* lipid accumulation process is a growth-associated process and does not require a nitrogen-limited medium (Papanikolaou and Aggelis, 2011). In particular, it is used for the biomodification of fats and oils to high-value polyunsaturated fatty acids.

### 2.2.5.1 Single Cell Oil Production Using Hydrophilic Feedstock

SCO production from food waste such as tomato waste, pear pomace, whey, and shrimp-processing by-products was investigated using various oleaginous fungi and yeasts. The yields achieved in recent SCO studies using hydrophilic substrates are summarized in Table 2.5. The highest SCO yield (1.35 g/g sugar) with a maximum γ-linoleic acid concentration of 301 mg/L was obtained from whey using *Mortierella isabellina*. Vamvakaki et al. (2010) also used whey after a physico-chemical pretreatment to produce SCO, and obtained a lipid content of 65% using *Cryptococcus curvatus* after 36 hours. They reported that the biomass and lipid yields were two times higher than those obtained using glucose, galactose, and lactose, and the lipid productivity (4.68 g/L/day) was the highest ever reported. These studies demonstrated that whey can be valorized through the production of biomass and SCO. Food wastes containing organic nitrogen, such as tomato waste also cause severe environmental problems. Fakas et al. (2008a) used tomato waste hydrolysate as a nitrogen source to produce SCO with high γ-linoleic acid concentration (800–1000 mg/L) using *Cunninghamella echinulata*. Microbial oil production from N-acetylglucosamine rich by-products of shrimp processing was also investigated using different oleaginous microorganisms (Zhang et al., 2011). Among them, *Cryptococcus curvatus* was the only lipid producer, and the highest lipid yield (0.35 g/g dry cell) was obtained in a nitrogen- and phosphorus-limited medium in stationary phase. Although the lipid biosynthesis is better after the growth phase, the rare polyunsaturated fatty acids, particularly γ-linoleic acid biosynthesis, is better in young and fast growing mycelia (Ratledge and Wynn, 2002; Fakas et al., 2009). Therefore, it is important to optimize the process conditions based on the targeted SCO application.

### 2.2.5.2 Single Cell Oil Production Using Hydrophobic Feedstock

Oil/fat wastes threaten the environment as they require a high energy demanding degradation process (Bialy et al., 2011). SCO production from waste fats/oils not only provides a smooth solution to this issue, but also adds value to waste oils by producing tailor-made lipids that can be used as cacao butter and/or medical/dietetic supplements (Papanikolaou et al., 2003). To date, SCO production from waste cooking oils, tallow, and animal fats was investigated using various oleaginous fungi and yeasts (Table 2.6). Among the studies, the highest SCO yield (0.88 g/g fat) with 58% lipid content was obtained from waste oil using *Yarrowia lipolytica* (Bialy et al., 2011), while the biomass with the highest lipid content (64%) was obtained from waste cooking olive oil using *Aspergillus* sp. ATHUM 3482 (Papanikolaou et al., 2011).

## TABLE 2.5
### SCO Production from Food Waste Using Hydrophilic Substrates

| Waste | Microorganism | Pretreatment | Vessel Type | Conditions | Duration (hours) | $Y_{biomass}$ (g/g waste) | $Y_{SCO}$ (g/g sugar) | $Y_{SCO}$ (g/g cell) | P (g/Lh) | References |
|---|---|---|---|---|---|---|---|---|---|---|
| Tomato waste hydrolyzate | Cunninghamella echinulata CCF-103 | 2% (v/v) $H_2SO_4$ treatment, autoclave (121°C, 2 h), filtration, pH adjustment to 6.0 | Flask | 28°C, pH 6, 180 rpm | 144 | NR | 0.25 | 0.40 | NR | Fakas et al. (2008b) |
| Tomato waste hydrolyzate | Cunninghamella echinulata ATHUM 4411 | 2% (v/v) $H_2SO_4$ treatment, autoclave (121°C, 2 h), filtration, pH adjustment to 6.0 | Flask | 28°C, pH 6, 180 rpm, DO:≥70% | 144 | NR | NR | 0.48 | NR | Fakas et al. (2008a) |
| Pear pomace | Mortierella isabellina | Homogenization | Petri dishes | 28°C, pH 6.5, 180 rpm, DO:≥70% | 212 | NR | NR | 0.12 | NR | Fakas et al. (2009) |
| Whey | Mortierella isabellina | NR | Flask | 28°C, pH 6, 180 rpm | 171 | NR | 0.55 | 0.17 | NR | Vamvakaki et al. (2010) |
| Whey | Thamnidium elegans | NR | Flask | 28°C, pH 6, 180 rpm | 219 | NR | 1.35 | 0.03 | NR | Vamvakaki et al. (2010) |
| Whey | Mucor sp. | NR | Flask | 28°C, pH 6, 180 rpm | 243 | NR | 1.02 | 0.09 | NR | Vamvakaki et al. (2010) |
| Whey | Cryptococcus curvatus | Hydrodynamic cavitation for 20 min at pH 9, pH adjustment to 5.5, filtration | Flask | 30°C, pH 5.5, 200 rpm | 36 | NR | NR | 0.65 | 4.68 | Seo et al. (2014) |
| Whey concentrate | Cunninghamella echinulata ATHUM 4411 | NR | Flask | 28°C, pH 6, 180 rpm | 144 | NR | 0.24 | 0.14 | NR | Fakas et al. (2008b) |
| By-products from shrimp processing | Cryptococcus curvattus | NR | Flask | 30°C, 115 rpm | 98 | NR | 0.28 | 0.35 | NR | Zhang et al. (2011) |

Source: Uçkun, E. K., A. P. Trzcinski, and Y. Liu: Platform chemicals production from food wastes using a biorefinery concept. Journal of Chemical Technology and Biotechnology. 2015. 90 (8). 1364–1379. Copyright Wiley-VCH Verlag GmbH & Co. KGaA. Reproduced with permission.

Note: AR: aeration rate, NR: not reported, fb: fed-batch, DO: dissolved oxygen.

**TABLE 2.6**
**SCO Production from FW Using Hydrophobic Substrates**

| Waste | Microorganism | Pretreatment | Vessel Type | Conditions | Duration (hours) | $Y_{biomass}$ (g/g fat) | $Y_{SCO}$ (g/g fat) | $Y_{SCO}$ (g/g cell) | P (g/Lh) | References |
|---|---|---|---|---|---|---|---|---|---|---|
| Animal fat | Yarrowia lipolytica | NR | Bioreactor with 1.25 L working vol. | 28°C, pH 6 | 120 | 1.10 | NR | 0.54 | NR | Papanikolaou et al. (2002) |
| Waste oil from chicken product fat | Yarrowia lipolytica (NC1) | NR | Flask | 28°C, pH 5.5 | 144 | 1.80 | 0.68 | 0.38 | NR | Bialy et al. (2011) |
| Waste oil from frying fish | Yarrowia lipolytica (NC1) | NR | Flask | 28°C, pH 5.5 | 144 | 1.68 | 0.76 | 0.45 | NR | Bialy et al. (2011) |
| Meat product fat | Yarrowia lipolytica (NC1) | NR | Flask | 28°C, pH 5.5 | 144 | 1.61 | 0.55 | 0.34 | NR | Bialy et al. (2011) |
| Waste oil from frying vegetables | Yarrowia lipolytica (NC1) | NR | Flask | 28°C, pH 5.5 | 144 | 1.51 | 0.88 | 0.58 | NR | Bialy et al. (2011) |
| Tallow | Yarrowia lipolytica ACA-DC 50109 | NR | Bioreactor with 1.25 L working vol. | 28°C, pH 6, 650 rpm, AR:0.3 vvm | 120 | 1.10 | NR | 0.52 | NR | Papanikolaou et al. (2007) |
| Waste cooking olive oil | Aspergillus sp. ATHUM 3482 | Filtration, homogenization | Flask | 28°C, pH 6, 200 rpm | 191 | 1.15 | 0.74 | 0.64 | NR | Papanikolaou et al. (2011) |
| FFA derived from animal and vegetable fats | Yarrowia lipolytica ACA-DC 50109 | NR | Bioreactor with 1.25 L working vol. | 28°C, pH 6, 300 rpm, AR:0.1 vvm | 130 | NR | 0.46 | NR | NR | Papanikolaou and Aggelis (2003) |

Source: Uçkun, E. K., A. P. Trzcinski, and Y. Liu: Platform chemicals production from food wastes using a biorefinery concept. *Journal of Chemical Technology and Biotechnology.* 2015. 90 (8). 1364–1379. Copyright Wiley-VCH Verlag GmbH & Co. KGaA. Reproduced with permission.

*Note:* AR: aeration rate, NR: not reported, fb: fed-batch, FFA: fatty acids.

These studies demonstrated that the fatty acid composition changed according to the type and concentration of the oil/fat used before and after fermentation. Bialy et al. (2011) reported that the oleic (C18:1) and linoleic acid (C18:2) contents increased from 42.38 and 7.89% to 50.48 and 16.63%, respectively, while the palmitic acid (C16:0) decreased from 41.88 to 19.87% after the fermentation of waste oil. After the fermentation of waste cooking oil, intracellular fatty acids were mainly composed of oleic acid (C18:1); therefore, the biomass and SCO produced was proposed to be used as an excellent food or feed nutritional supplement (Papanikolaou et al., 2011). Papanikolaou and Aggelis (2003) demonstrated that different fatty acids are selectively removed and/or accumulated in different stages of the microbial growth. At late fermentation steps, cellular lipid degradation increased due to the decreased uptake rate of the fat. In order to accumulate the desired lipids, a fed-batch strategy was suggested (Papanikolaou et al., 2011).

The utilization of both hydrophilic and hydrophobic waste to produce SCO is an alternative approach to treat waste and produce high-value products. The process costs may be high, but the utilization of low/no-cost food waste may result in an affordable process. Moreover, the recent developments in the area of lipid biotechnology can further improve the SCO production, novel fatty acid synthesis, and lipid alteration. Currently, there is only one pilot plant producing SCO from waste using oleaginous yeasts—microalgae. The commercialization of the produced SCO is reported to be in 5–10 years at the earliest (Neste-oil, 2012).

## 2.3  OTHER CHEMICALS

Besides the chemicals mentioned above, there are many other chemicals and biopolymers that can be produced via fermentation. These include malic, fumaric, acetic, propionic, and butyric acids that have growing markets. There are a few studies reporting high concentrations of these acids using pure glucose, molasses, starch hydrolyzates, or glycerol with pure cultures, but studies on mixed FW are still rare. A serious obstacle, however, is the contamination with other compounds in FW which may render the downstream purification particularly difficult or the production of toxins which may disqualify these chemicals in the food industry. Strain selection or genetic improvement may help to avoid the production of these toxins. Other than these common organic acids, 1,3-propanediol, 3-hydroxypropionic acid, and 2,3-butanediol also have a great potential to be produced from FW via fermentative processes.

## 2.4  CONCLUSIONS

In conclusion, large amounts of FW are generated worldwide. Environmental concerns directed the research for alternative, environmentally friendly methods to handle FW. The publications discussed above on the valorization of FW using biological treatment strategies indicated that a wide range of platform chemicals can be produced from FW. Readily available FW, which had previously been considered of low or no economic value, is recently being recognized as raw materials with potential value to produce platform chemicals via fermentation. However, most of the valorization techniques discussed above have been validated only in the laboratory, and only a few were carried out at pilot scale. Further optimization and scale-up studies need to be carried out in order to exploit FW for the production of platform chemicals. The utilization of engineered strains and process integration techniques would provide a better conversion of waste and a more economical biorefinery. The efficiency and cost base of the production could be further improved by integrating different value-added product processes. However, the initial costs of such an integrated biorefinery system would be high because this new approach is still immature and needs to be developed further. Still, the low or no cost of food waste along with the environmental benefits and waste disposal costs could balance the initial high capital costs of the biorefineries.

# 3 Enzyme Production from Food Wastes

## 3.1 INTRODUCTION

Food wastes can be dumped, landfilled, incinerated, composted, digested anaerobically, and used as animal feed. In many Asian countries, FW is still dumped with other household waste in landfills or dumpsites (Figure 3.1). Unfortunately, the capacity of landfills is limited and environmental pollution (leachate, gas, odors, flies, vermin, and pathogens) poses serious problems (Ngoc and Schnitzer, 2009). Hence, there is a need for an appropriate management of FW (Ma et al., 2009a). In order to reduce its volume, FW is traditionally incinerated with other combustible municipal wastes for the production of heat or energy. The incineration of FW is a favorable option against landfilling in terms of overall energy recovery, and emissions of greenhouse gases (Othman et al., 2013). However, the incineration may not be an affordable approach for most low income countries due to the high capital and operating costs (Ngoc and Schnitzer, 2009). Moreover, incineration of FW can potentially cause air pollution (Takata et al., 2012).

Another approach to handling biodegradable FW is composting, which results in a valuable soil conditioner and fertilizer (Gajalakshmi and Abbasi, 2008). Composting facilities showed a relatively low environmental impact and a high economic efficiency compared to other treatment methods. The primary recycling method in Korea is composting (Figure 3.1). However, the high moisture content of FW causes remarkable levels of leachate which affects process performance by reducing oxygen availability and weakening the pile strength (Cekmecelioglu et al., 2005). In this case, high airflows for aeration or excessive carbon ingredients are necessary for process control, which increases operational costs. Indeed, compost is more expensive than commercial fertilizers and the available market for compost is not large (Aye and Widjaya, 2006).

Anaerobic digestion is another alternative which yields methane and carbon dioxide as metabolic end products, and is therefore feasible from an economic and environmental point of view because methane is used as an energy source (Othman et al., 2013). Hirai et al. (2001) evaluated the environmental impacts of FW treatment and found that utilizing a methane fermentation process prior to incineration reduces approximately 70 kg $CO_2$ eq/tonne waste of the global warming potential due to the substitution effect. The disadvantages of using FW as animal feed are the variable composition and the high moisture content, which favors microbial contamination (Esteban et al., 2007). To prevent this, animal feed is generally dried but greenhouse gas emission increases depending on the energy usage during the drying process, which is related to the water content of FW (Takata et al., 2012).

FW is mainly composed of carbohydrate polymers (starch, cellulose, and hemicelluloses), lignin, proteins, lipids, and organic acids (Table 3.1). Total sugar and

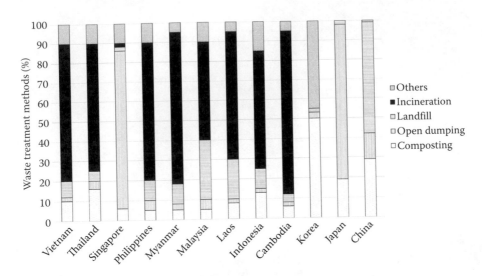

**FIGURE 3.1** Waste treatment methods in some Asia–Pacific countries. (With kind permission from Springer Science+Business Media: *Waste and Biomass Valorization*, Enzyme production from food wastes using a biorefinery concept: A review, 5, 2014a, 903–917, Uçkun, K.E. et al.)

protein contents in FW are in the range of 35.5%–69% and 3.9%–21.9%, respectively. Due to its inherent chemical complexity, alternative treatment methods are currently studied and attention is being directed to production of high value-added products such as biofuels, biodiesel, platform chemicals, and enzymes (Han and Shin, 2004b; Sakai and Ezaki, 2006; Yang et al., 2006; Pan et al., 2008; Zhang et al., 2010; He et al., 2012a,b; Zhang et al., 2013b;Vavouraki et al., 2014). As a comparison, fuel applications ($200–400/ton biomass) and organic acids, biodegradable plastics, and enzymes applications ($1000/ton biomass) usually create more value compared to generating electricity ($60–150/ton biomass) and animal feed ($70–200/ton biomass) (Sanders et al., 2007).

The critical stage of biomass bioconversion is saccharification, which hampers its commercial use. For an efficient biomass conversion, carbohydrate components of FW should be hydrolyzed to yield high concentrations of oligosaccharides and monosaccharides, which are amenable to fermentation. Hence, there is an increasing interest on the production of biomass saccharifying enzymes, mainly amylases and cellulases (Teeri, 1997). There are remarkable amounts of publications on the lab-scale production of various industrial enzymes such as proteases, amylases, ligno-cellulosic enzymes, and lipases using different types of FW. Therefore, this chapter summarizes and discusses recent enzyme production studies from FW.

## 3.2  ENZYME PRODUCTION

Enzymes are commonly used in many industrial applications due to their great selectivity for the substrates and their biodegradabilities. They act under mild and

**TABLE 3.1**

**Characteristics of Mixed Food Waste**

| Origin | pH | Moisture | Total Solid | VS/TS | Total Sugar | Starch | Cellulose | Lipid | Protein | Ash | References |
|---|---|---|---|---|---|---|---|---|---|---|---|
| Dining hall | NR | 79.5 | 20.5 | 95.0 | NR | NR | NR | NR | 21.9 | NR | Han and Shin (2004b) |
| Cafeteria | 5.1 | 84.1 | 15.9 | 15.2 | NR | NR | NR | NR | NR | NR | Kim et al. (2008c) |
| Cafeteria | 5.1 | 80.0 | 20.0 | 93.6 | NR | NR | NR | NR | NR | 1.3 | Kwon and Lee (2004) |
| MSW | NR | 85.0 | 15.0 | 88.5 | NR | NR | 15.5 | 8.5 | 6.9 | 11.5 | Rao and Singh (2004) |
| Cafeteria | 4.6–5 | 79.1 | 20.9 | 93.2 | NR | NR | NR | NR | NR | NR | Ramos et al. (2012) |
| Cafeteria | NR | 75.9 | 24.1 | NR | 42.3 | 29.3 | NR | NR | 3.9 | 1.3 | Ohkouchi and Inoue (2006) |
| NR | NR | 87.6 | 12.4 | 89.3 | NR | NR | NR | NR | NR | NR | Kim et al. (2006b) |
| Residents | 4.9 | 80.8 | 19.2 | 92.7 | NR | 15.6 | NR | NR | NR | NR | Pan et al. (2008) |
| Dining hall | NR | 80.3 | 19.7 | 95.4 | 59.8 | NR | 1.6 | 15.7 | 21.8 | 1.9 | Tang et al. (2008) |
| Dining hall | NR | 82.8 | 17.2 | 89.1 | 62.7 | 46.1 | 2.3 | 18.1 | 15.6 | NR | Wang et al. (2008c) |
| Restaurant | 3.9 | 80.0 | 20.0 | 95.0 | 70.0 | NR | NR | 10.0 | 13.0 | NR | Zhang et al. (2005) |
| Dining hall | 5.6 | 82.8 | 17.2 | 85.0 | 62.7 | 46.1 | 2.3 | 18.1 | 15.6 | NR | Ma et al. (2008) |
| Cafeteria | NR | 61.3 | 38.7 | NR | 69.0 | NR | NR | 6.4 | 4.4 | 1.2 | Uncu and Cekmecelioglu (2011) |
| Food court | NR | 64.4 | 35.6 | NR | NR | NR | NR | 8.8 | 4.5 | 1.8 | Cekmecelioglu and Uncu (2013) |
| Canteen | NR | 81.7 | 18.3 | 87.5 | 35.5 | NR | NR | 24.1 | 14.4 | NR | He et al. (2012a) |
| Restaurant | NR | 81.5 | 18.5 | 94.1 | 55.0 | 24.0 | 16.9 | 14.0 | 16.9 | 5.9 | Vavouraki et al. (2014) |
| Restaurant | NR | 81.9 | 14.3 | 98.2 | 48.3 | 42.3 | NR | NR | 17.8 | NR | Zhang and Jahng (2012) |

*Source:* With kind permission from Springer Science+Business Media: *Waste and Biomass Valorization*, Enzyme production from food wastes using a biorefinery concept: A review, 5, 2014a, 903–917, Uckun, K.E. et al.

*Note:* Total solid, total sugar, starch, cellulose, lipid, protein and ash contents are given in wt% on the basis of dry weight. Volatile solid contents are given as the %VS ratio on total solid basis. NR: not reported.

environmentally friendly conditions. Hence, enzyme production is one of the most important applications, which serves various industries. Research is continuing on the production of different enzymes in solid state fermentation (SSF) with the ultimate aims to obtain high activity enzymes at lower cost using low cost substrates and/or by improving economical processing technologies. There are remarkable amounts of publications on the production of various enzymes using different agro-industrial waste (Prakasham et al., 2005; Chutmanop et al., 2008; De Castro et al., 2011; Vaseghi et al., 2013). However, the main problem is the recalcitrant nature, which results in low enzyme yields and expensive enzyme production. The recalcitrant nature can be mitigated by some pretreatment steps while enzyme yields can be enhanced by developing suitable fermentation conditions or using genetically modified microbial strains (Pandey et al., 2000b). On the other hand, enzyme production costs can be reduced by developing suitable fermentation processes using FW, which have easily digestible components. There are some publications reporting the production of enzymes from FW using both solid and submerged fermentation systems (Tables 3.2 through 3.6). Various kinds of FW were used to produce enzymes such as proteases, cellulases, amylases, lipases, and pectinases, particularly using SSF. SSF has several advantages over submerged fermentation (SmF) as it requires less capital investment, lower energy, and a simple fermentation medium; it has superior productivity and produces less wastewater (Couto and Sanromán, 2006). Moreover, an easy control of bacterial contamination and lower costs of downstream processing make it more attractive. It is appropriate for the production of enzymes, especially because of the higher enzyme yields that can be obtained compared to submerged fermentation (Shukla and Kar, 2006; Ruiz et al., 2012; Thomas et al., 2013). SSF provides a similar environment to the microorganism's natural environment, which provides better conditions for its growth and enzyme production (Thomas et al., 2013). However, there are only a few reports on SSF bioreactor design in the literature. The large-scale production of enzymes using SSF is challenging because pH, temperature, aeration, oxygen transfer, and moisture content is difficult to control (Couto and Sanromán, 2006; Wang et al., 2008b).

## 3.2.1 AMYLASES

The amylase family has two major classes, namely α-amylase (EC 3.2.1.1) and glucoamylase (GA) (EC 3.2.1.3). α-amylase hydrolyses starch into maltose, glucose, and maltotriose by cleaving the 1,4-α-D-glucosidic linkages between adjacent glucose units in the linear amylose chain (Pandey et al., 2000a) while glucoamylase hydrolyses the non-reducing ends of amylose and amylopectin to glucose (Anto et al., 2006). Amylases have been widely used in the food, fermentation, textiles, and paper industries (Pandey et al., 2000a). They are also used for the pretreatment of the agro-industrial and organic by-products to improve the bioproduct yield in subsequent processes. There is an increasing interest in the production of amylases using cheap feedstocks (Wang et al., 2008b). High activity amylases can be produced from FW such as kitchen refuse (Wang et al., 2008b), potato peel (Shukla and Kar, 2006; Elayaraja et al., 2011), coffee waste (Murthy et al., 2009), and tomato pomace (Umsza-Guez et al., 2011) via the optimization of fermentation using different

**TABLE 3.2**

**Amylase Production from Food Wastes**

| Food Waste | Microorganism | Pretreatment Method | Fermentation Mode and Vessel Type | Fermentation Conditions | Duration (day) | Achievements | References |
|---|---|---|---|---|---|---|---|
| Potato peel | *Bacillus subtilis* | Dried, ground, sieved | SSF-250 mL flasks | 40°C, pH 7, 65% MC, 10% (v/w) inoculum | 2 | α-amylase (600 U/mL) | Shukla and Kar (2006) |
| Potato peel | *Bacillus licheniformis* | Dried, ground, sieved | SSF-250 mL flasks | 40°C, pH 7, 70% MC, 10% (v/w) inoculum | 2 | α-amylase (270 U/mL) | Shukla and Kar (2006) |
| Coffee waste | *Neurospora crassa* CFR 308 | Ground, steamed | SSF-250 mL flasks | 28°C, pH 4.6, 60% MC, 1 mm PS, $10^7$ spores/g ds, | 5 | α-amylase (6342 U/g ds) | Murthy et al. (2009) |
| Potato peel | *Bacillus firmus* CAS 7 | Dried, ground, sieved | SmF-250 mL flasks | 35°C, pH 7.5, 1% S | 2 | α-amylase (676 U/mL) | Elayaraja et al. (2011) |
| Tomato pomace | *Aspergillus awamori* | Dried, milled, sieved | SSF-plate-type SSF bioreactor | 28°C, pH 5 | 5 | α-amylase (10.9 IU/g ds) | Umsza-Guez et al. (2011) |
| Bread waste | *Bacillus caldolyticus* DSM 405 | NR | SmF-1 L flask with 100 mL working vol. | 30°C, pH 7 | 1 | α-amylase (6.7 U/g ds) | Jamrath et al. (2012) |
| Pea pulp | *Bacillus caldolyticus* DSM 405 | None | SmF flasks | 70°C, 150 rpm | 6 | α-amylase (8.6 U/mL) | Jamrath et al. (2012) |
| Food waste | *Aspergillus niger* UV-60 | None | SmF-250 mL flasks | 30°C, pH 5, 120 rpm, 5% I/S | 4 | GA (137 U/mL) | Wang et al. (2008b) |
| Bread waste | *Aspergillus oryzae* | None | SSF Petri plates | 30°C, MC:1.8 (w/w, db), PS:20 mm, $10^6$ spore/gdS | 6 | GA (114 U/gdS) | Melikoglu et al. (2013c) |

*Source:* With kind permission from Springer Science+Business Media: *Waste and Biomass Valorization*, Enzyme production from food wastes using a biorefinery concept: A review, 5, 2014a, 903–917, Uçkun, K.E. et al.

*Note:* S: substrate, SSF: solid state fermentation, SmF: submerged fermentation, I/S: inoculum to substrate ratio, ds: dry substrate, MC: moisture content, PS: particle size, ds: dry solid, GA: glucoamylase.

microbial strains. However, it is not easy to compare the efficiency of the processes as the produced enzymes' activities are defined differently (Table 3.2). The main advantages of FW utilization for enzyme production are that fermentations do not require harsh pretreatments and nutrient addition.

Wang et al. (2008b) investigated the production of glucoamylase from FW by *Aspergillus niger* UV-60 using SmF. They reported that nutrients supplementation including yeast extract, $(NH_4)_2SO_4$, $KH_2PO_4$, $MgSO_4 \cdot 7H_2O$, $FeSO_4 \cdot 7H_2O$, $CaCl_2$, and particle size reduction had no significant influence on the glucoamylase production. Maximum glucoamylase activity of 137 U/mL was obtained using 3.75% FW and 5% (v/w, $10^6$ spores/mL) inoculum at 30°C, 120 rpm for 96 h. A reducing sugar concentration of 60.1 g/L was produced from 10% FW (w/v), within 125 minutes using the produced crude glucoamylase. Shukla and Kar (2006) produced high activity α-amylase from potato peels by SSF using two thermophilic isolates of *Bacillus licheniformis* and *Bacillus subtilis*. Under optimal conditions (40°C, pH 7, 1 mm particle size with 65%–70% moisture content), α-amylase activities obtained using *B. licheniformis* and *B. subtilis* were 270 and 600 U/mL, respectively. In another study, α-amylase production from potato peels was conducted by SmF using a thermophilic isolate of alkaline tolerant *B. firmus* CAS7 strain (Elayaraja et al., 2011). Under the optimal conditions (at 35°C, pH 7.5 and 1% substrate), 676 U/mL of α-amylase which was optimally active at 50°C and pH 9 was obtained. Murthy et al. (2009) used coffee wastes as the sole carbon source for the synthesis of α-amylase in SSF using a fungal strain of *Neurospora crassa* CFR 308. α-amylase activity of 4324 U/g dry substrate was obtained using 1 mm particle size, $10^7$ spores/g dry substrate, 60% moisture content at 28°C, pH 4.6. Steam pretreatment improved the accessibility of coffee waste and an α-amylase activity of 6342 U/g dry substrate was obtained.

FW can be used to produce high activity amylases using suitable microbial strains. In some of the lactic acid production studies from FW, a saccharification step using commercial amylases was conducted prior to the fermentation in order to improve and ease the fermentation process (Kim et al., 2003; Sakai et al., 2004b). If the enzyme production step can be integrated to the fermentation system, the process costs could be lowered. In a study by Leung et al. (2012), bread waste was used as the sole feedstock in a biorefinery concept for the production of succinic acid (SA), one of the future platform chemicals of a sustainable chemical industry. Bread waste was used in the SSF of *Aspergillus awamori* and *Aspergillus oryzae* to produce enzyme complexes rich in amylolytic and proteolytic enzymes. The resulting fermentation solids were added directly to a bread suspension to generate a hydrolysate rich in glucose and free amino nitrogen. The bread hydrolyzate was used as the sole feedstock for *A. succinogenes* fermentations, which led to the production of 47.3 g/L succinic acid with 1.16 g SA/g glucose yield, which is the highest succinic acid yield compared from other FW-derived media reported to date. This consolidated process could be potentially utilized to transform no-value FW into succinic acid.

### 3.2.2 Lignocellulolytic Enzymes

Lignocellulolytic enzymes are mainly produced by several fungi, and are composed of cellulases, xylanases, and ligninases, which degrade lignocellulosic materials.

Cellulases have many applications in various industries including food, animal feed, brewing and wine making, agriculture, biomass refining, pulp and paper, textile, and laundry (Kuhad et al., 2011). The bioconversion of cellulose to fermentable sugars requires the synergistic action of three enzyme classes: endoglucanases (EC 3.2.1.4), which act randomly on soluble and insoluble cellulose chains, exoglucanases (cellobiohydrolases; EC 3.2.1.91), which liberate cellobiose from the reducing and non-reducing ends of cellulose chains, and β-glucosidases (EC 3.2.1.21), which liberate glucose from cellobiose (Jørgensen et al., 2007). Xylanases have many applications in the food, feed, pulp and paper, brewing, wine making, and textile industries with or without concomitant use of cellulases (Khandeparkar and Bhosle, 2006). The hydrolysis of xylans mainly requires the action of endo-β-1,4-xylanase and β-xylosidase. However, the presence of other accessory enzymes is needed to hydrolyze substituted xylans (Uçkun et al., 2013a). Lignin is an undesirable polymer for biofuel production as it prevents the accessibility of plant-derived polysaccharides. However, lignin-derived materials can be used to develop valuable products such as dispersants, detergents, drilling mud thinner, surfactants, coagulants and flocculants (for sewage and wastewater treatment), adhesives, graft polymers including polyurethanes, polyesters, polyamines, epoxies, and rubbers (Effendi et al., 2008; Menon and Rao, 2012). In order to degrade lignin polymers, ligninolytic enzymes including laccases, lignin peroxidases, and Mn-peroxidase are used.

Lignocellulolytic enzymes are also used for the pretreatment of agro-industrial and organic by-products to improve the bioproduct yields in subsequent processes (Bansal et al., 2012; Saravanan et al., 2012). Recent studies on lignocellulosic enzyme production using FW and the achieved enzyme activities are summarized in Table 3.3. Since the definition of enzyme activity are different in each study, it is not an easy task to compare the achievements and detect the best method. However, generally, fungal SSF is the most preferred method due to its advantages over SmF (Krishna, 1999; Sun et al., 2010; Sun et al., 2011; Bansal et al., 2012; Dhillon et al., 2012; Saravanan et al., 2012). Krishna (1999) reported that the total cellulase production from banana waste was 12-fold higher in SSF than that obtained using SmF. However, Díaz et al. (2012) reported that the SmF resulted in higher xylanase production in comparison to SSF due to better aeration. Umsza-Guez et al. (2011) demonstrated a clear positive effect of aeration on xylanase and carboxymethyl cellulase (CMCase) production using SSF in a plate-type bioreactor.

The effects of process parameters such as incubation temperature, pH, moisture content, particle size, nutrients supplementation, inoculum size, and pretreatment methods on enzyme production have been investigated. In general, the optimum conditions in SSF depend not only on the microorganism employed, but also greatly on the type of substrate. The incubation time, pH, temperature, particle size, and water content should be optimized once the substrate and microorganisms are chosen. Some FW require extra nutrients (Sun et al., 2010; Umsza-Guez et al., 2011; Dhillon et al., 2012), while some others can be used as the sole source to produce high titers of cellulases (Botella et al., 2005; Sun et al., 2011; Bansal et al., 2012). Dhillon et al. (2012) analyzed the effects of inducers on cellulase and hemicellulase production by *Aspergillus niger* NRRL-567 using apple pomace as a substrate. The higher filter paper cellulase (FPA) and β-glucosidase activities of 133.68 ± 5.44 IU/gram dry

**TABLE 3.3**

**Lignocellulosic Enzyme Production from Food Wastes**

| Food Waste | Microorganism | Pretreatment | Fermentation Mode and Vessel Type | Fermentation Conditions | Duration (day) | Achievements | References |
|---|---|---|---|---|---|---|---|
| Banana wastes | Bacillus subtilis (CBTK106) | Dried, ground, acid and alkali pretreatment | SSF-250 mL flasks | 35°C, pH 7, 400 μm PS, 70% MC, 15% (v/w) I/S ratio | 3 | FPAse (2.8 IU/ds), CMCase (9.6 IU/g ds), Cellobiase (4.5 IU/g ds) | Krishna (1999) |
| Grape pomace | Aspergillus awamori | Dried, milled, sieved | SSF Petri dishes | 30°C, 10 g S, $5 \times 10^5$ I/S, 60% MC | 7 | Xylanase (40.4 IU/g ds), Cellulase (9.6 IU/g ds) | Botella et al. (2005) |
| Apple pomace | Trichoderma sp. | Dried, crushed, sieved | SSF-250 mL flasks | 32°C, 70% MC, $10^8$ spores/flask | 6 | Cellulase (5.8 U/g ds) | Sun et al. (2010) |
| Banana peel | Trichoderma viride GIM 3.0010 | Dried, crushed, sieved | SSF-250 mL flasks | 30°C, 65% MC, $10^9$ spores/flask | 6 | FPA(5.6U/g ds), CMCase (10.3 U/g ds), β-glucosidase (3U/g ds) | Sun et al. (2011) |
| Tomato pomace | Aspergillus awamori | Dried, milled, sieved | SSF-plate-type SSF bioreactor | 28°C, pH 5 | 5 | Xylanase (195.9 IU/g ds), CMCase (19.7 IU/g ds) | Umsza-Guez et al. (2011) |
| Carrot, orange, pineapple, potato peels, wheat bran | Aspergillus niger NS-2 | Acid/base pretreatment | SSF-250 mL flasks | 30°C, pH 7, 1:1.5 to 1:1.75 S/M ratio | 4 | CMCase (310 U/gds), FPase (17 U/gds), β-glucosidase (33 U/gds) using alkaline pretreated wheat bran | Bansal et al. (2012) |
| Apple pomace | Aspergillus niger NRRL-567 | Drying, crushing, sieving | SSF-500 mL flasks | 30°C, 1.7–2 mm PS, 75% MC, $10^7$ spores/g dS | 7 | FPase (113.7 IU/gds), CMCase (172.31 IU/gds), β-glucosidase (60.1IU/gds), Xylanase (1412.6 IU/gds) | Dhillon et al. (2012) |

*(Continued)*

**TABLE 3.3 (*Continued*)**
**Lignocellulosic Enzyme Production from Food Wastes**

| Food Waste | Microorganism | Pretreatment | Fermentation Mode and Vessel Type | Fermentation Conditions | Duration (day) | Achievements | References |
|---|---|---|---|---|---|---|---|
| Grape pomace and orange peel | *Aspergillus awamori* | Dried, milled and sieved | SSF Petri dishes | 30°C, pH 5, 70% MC, 4 5 × 10$^8$ spores/g S. | 15 | Exo-PG (3.8 IU/gds), Xylanase (32.7 IU/gds), Cellulase (5.4 IU/gds) | Díaz et al. (2012) |
| Potato peel | *Aspergillus niger* | Dried, ground | SSF | 30°C, 10$^7$ spores/g dS, 50% MC | 3 | FPase (0.015 U/mL), CMCase (0.023 U/mL), Xylanase(0.024 U/mL) | Dos Santos et al. (2012) |
| Mango peel | *Trichoderma reesei* | Alkaline pretreatment | SmF-250 mL flasks | 30°C, pH 7, 200 rpm | 6 | Cellulase (7.8 IU/mL) | Saravanan et al. (2012) |
| Passion fruit waste | *Pleurotus pulmonarius* | Dried, milled | SSF-250 mL flasks | 28°C in complete darkness | 14 | MnP (0.22 U/mL), β-xylosidase (4.76 U/mL), β-Glucosidase (2.96 U/mL), β-galactosidase (6.21 U/mL) | Zilly et al. (2012) |
| Passion fruit waste | *Macrocybe titans* | Dried, milled | SSF-250 mL flasks | 28°C in complete darkness | 14 | Laccase (10.2 U/mL), Pectinase (1.72 U/mL), Endoxylanase (0.27 U/mL) | Zilly et al. (2012) |

*Source:* With kind permission from Springer Science+Business Media: *Waste and Biomass Valorization*, Enzyme production from food wastes using a biorefinery concept: A review, 5, 2014a, 903–917, Uçkun, K.E. et al.

*Note:* S: substrate, SSF: solid state fermentation, SmF: submerged fermentation, I/S: inoculum to substrate ratio, DS: dry substrate, S/M: substrate to moisture ratio, MC: moisture content, PS: particle size, ds: dry solid, PG: polygalacturonase, CMCase: carboxymethylcellulase, MnP: manganese peroxidase, NR: not reported.

substrate (gds) and $60.09 \pm 3.43$ IU/gds, respectively, were observed while using $CuSO_4$ and veratryl alcohol. Similarly, higher xylanase activity of $1412.58 \pm 27.9$ IU/gds was observed with veratryl alcohol after 72 h of fermentation time, while a higher CMCase activity of $172.31 \pm 14.21$ IU/g ds was obtained with lactose after 48 hours of incubation. Sun et al. (2010) also reported that cellulase production using SSF was markedly improved by supplementing lactose and corn-steep solid to the apple pomace.

The effects of nutrients and other process parameters on cellulase production from banana waste by *Bacillus subtilis* (CBTK 106) was also evaluated by Krishna (1999). The optimal FPAse of 2.8 IU/g dry substrate, CMCase activity of 9.6 IU/g dry substrate, and cellobiase activity of 4.5 IU/g dry substrate were obtained at 72 hours incubation with media containing heat pretreated banana fruit stalk, $(NH_4)_2SO_4$, $NaNO_3$, and glucose. Saravanan et al. (2012) investigated the cellulase production from mango peel using *Trichoderma reesei* and reported that avicel, soybean cake flour, $KH_2PO_4$, and $CoCl_2 \cdot 6H_2O$ have a positive influence on cellulase production. Cellulase activity was 7.8 IU/mL using the optimum nutrient concentrations of 25.3 g/L avicel, 23.53 g/L soybean cake flour, 4.9 g/L $KH_2PO_4$, and 0.95 g/L $CoCl_2$ $6H_2O$, which was determined by response surface methodology.

Díaz et al. (2012) reported that cellulase production was inhibited at high concentration of reducing sugars when grape pomace was used as a substrate. They avoided this problem by adjusting the nutrient composition of grape pomace by supplementing orange peel, which is a pectin, cellulose- and hemicellulose-rich substrate inducing cellulase production. The synthesis of xylanase and cellullase increased using the mixed type substrate compared to whole grape pomace. Umsza-Guez et al. (2011) reported that xylanase production from tomato wastes using SSF was activated by $Mg^{2+}$, but is strongly inhibited by $Hg^{2+}$ and $Cu^{2+}$.

The effects of substrate pretreatments on cellulase and xylanase production have been studied (Krishna, 1999; Saravanan et al., 2012). Bansal et al. (2012) studied the effects of acid and base pretreatment on cellulase production from FW including carrot peelings, orange peelings, pineapple peelings, potato peelings, and wheat bran using SSF. The pretreated substrates are well suited for the organism's growth, producing high titers of cellulases after 96 h without the supplementation of nutrients. Yields of cellulase were higher in alkali-treated substrates compared to acid-treated and untreated substrates except in wheat bran. Of all the substrates tested, untreated wheat bran induced the maximum production of enzyme components followed by alkali-treated composite kitchen waste and potato peelings. Krishna (1999) investigated the effects of acid, alkaline, and heat pretreatment on cellulase production from banana waste using *Bacillus subtilis*. Although cellulase production was not affected by alkali or acid treatment, it increased 6.84-fold using a pressure-cooking method under controlled pH. Pressure cooking of plant materials at a controlled pH resulted in greater substrate accessibility for microbial growth. Moreover, it did not result in the formation of monosaccharide degradation products, such as furfural and hydroxymethyl furfural, which otherwise inhibit the cellulases (Weil et al., 1994).

Besides cellulases and xylanases, ligninases were also produced from FW by white rot fungi. Zilly et al. (2012) studied oxidative and hydrolytic enzyme production by SSF of yellow passion fruit waste using the white-rot fungi *Pleurotus ostreatus, Pleurotus pulmonarius, Macrocybe titans, Ganoderma lucidum,* and *Grifola frondosa.* Under the conditions used, the main enzymes produced were laccases, pectinases, and aryl-β-D-glycosidases (β-glucosidases, β-xylosidases, and β-galactosidases). The yellow passion fruit waste was as good as wheat bran, which is the most commonly used substrate for white-rot fungi cultivation.

Biorefineries need to develop their indigenous enzyme production processes along with their existing processes (Figure 3.2) as commercial enzyme production systems are still expensive to incorporate in biorefineries (Chandel et al., 2012). As can be seen from the studies above, some strains are producing different lignocellulosic

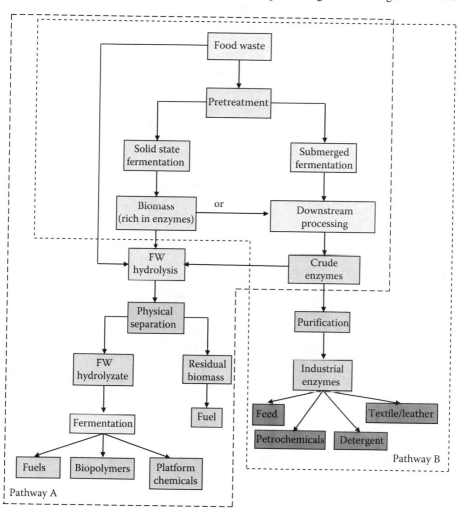

**FIGURE 3.2**  Simplified process flow diagram for an FW-based biorefinery concept. (With kind permission from Springer Science+Business *Media: Waste and Biomass Valorization,* Enzyme production from food wastes using a biorefinery concept: A review, 5, 2014a, 903–917, Uçkun, K.E. et al.)

enzymes from food wastes simultaneously. These enzyme cocktails can be used to hydrolyze biomass effectively at a low cost for their conversion to biofuels, platform chemicals, and biodegradable polymers. To further improve the hydrolysis, different strains can be used to produce enzyme solutions with different hydrolytic activities. Besides, some engineered strains can be used to improve the saccharification yield.

### 3.2.3 PECTINOLYTIC ENZYMES

Pectinolytic enzymes are enzymes degrading pectin polymers in a sequential and synergic way, by depolymerization and deesterification reactions. Complete degradation of pectin requires endo- and exo-acting polygalacturonases, pectin and pectate lyases as well as enzymes that cleave the rhamnogalacturonan chain, the rhamnogalacturonases (Kashyap et al., 2001). Pectinases are widely used in the food industry particularly for juice and wine production and many other conventional industrial processes, such as textile, plant fiber processing, tea, coffee, oil extraction, and treatment of industrial wastewater (Botella et al., 2007; Pedrolli et al., 2009; Ruiz et al., 2012). The production of pectinases is mainly conducted via fungal SSF particularly by using *Aspergillus* strains (Kashyap et al., 2001). For industrial implementation, pectinases can be produced from pectin-containing wastes, such as citrus and orange wastes (Garzon and Hours, 1992; Giese et al., 2008; Afifi, 2011), apple pomace (Hours et al., 1988; Berovic and Ostroversnik, 1997), grape pomace (Botella et al., 2005), and many other fruit residues (Martínez Sabajanes et al., 2012) without any harsh pretreatment owing to the nature of these substrates and the low moisture content (Botella et al., 2007; Martínez Sabajanes et al., 2012). Hours et al. (1988) investigated pectinase production from apple pomace by SSF using *Aspergillus foetidus*. The medium composition, temperature, and type of apple pomace used affected the enzyme production. After 36 hours of culture at 30°C with organic nitrogen supplemented to the apple pomace medium, an enzyme activity of 1300 U/g was obtained (Table 3.4).

In another study, pectinolytic enzyme production from citrus waste was studied using *Aspergillus foetidus* (Garzon and Hours, 1992). Yeast extract and mineral salt addition improved the activity up to 1600–1700 U/g after 36 hours of SSF. Berovic and Ostroversnik (1997) reported that pectolytic enzyme production from apple pomace using SSF with *Aspergillus niger* was induced and/or improved by supplementing the media with other cheap nutrients such as soya flour, wheat bran, wheat corn, and whey. They also mentioned that the highest activity was obtained using 38% moisture content and moisture content is very important in enzyme production. However, Ruiz et al. (2012) reported that 70% moisture content gave the highest pectinase activity using lemon peel pomace. Botella et al. (2007) evaluated the feasibility of grape pomace for the production of exo-polygalacturonase by *Aspergillus awamori* in SSF. The particle size of the substrate did not influence the enzyme production as it was reported by Hours et al. (1988), while the addition of extra carbon sources and the initial moisture content of the grape pomace were found to have a marked influence on the enzyme yield. In another study, Giese et al. (2008) carried out the production of pectinases from orange waste by *Botryosphaeria rhodina* MAMB-05 using both SSF and SmF with and without adding nutrients. Orange

## TABLE 3.4
## Pectinolytic Enzyme Production from Food Wastes

| Food Waste | Microorganism | Pretreatment Method | Fermentation Mode and Vessel Type | Fermentation Conditions | Duration (day) | Achievements | References |
|---|---|---|---|---|---|---|---|
| Apple pomace | *Aspergillus foetidus* NRRL 341 | None | SSF Petri dishes | 30°C, pH 4, $10^3$ 1/S | 2 | Pectinase (1300 U/g S) | Hours et al. (1988) |
| Citrus waste | *Aspergillus foetidus* NRRL 341 | None | SSF Petri dishes | 30°C | 2 | Pectinase (1641 U/g S) | Garzon and Hours (1992) |
| Apple pomace | *Aspergillus niger* | None | SSF-15 L horizontal solid state stirred tank reactor | 35°C | 3 | 900 AJDA U/mL | Berovic and Ostroversnik (1997) |
| Grape pomace | *Aspergillus awamori* | Milled, sieved | SSF Petri dishes | 30°C, 60% MC | 1 | Exo-PG(40U/g S), Xylanase (40 U/g S) | Botella et al. (2007) |
| Orange bagasse | *Botryosphaeria rhodina* MAMB-05 | Dried, ground | SSF-125 mL flask | 28°C | 6 | Pectinase (32 U/mL), Laccase (46 U/mL) | Giese et al. (2008) |
| Orange waste | *Aspergillus giganteus* CCT3232 | NR | SmF flask | 30°C, pH 6, 120 rpm, $10^7$ spores/mL | 3.5 | Exo-PG (48.5 U/mL) | Pedrolli et al. (2008) |
| Fruit residues (apple, lemon peel, grape skin, and tamarind kernel) | *Aspergillus flavipes* FP-500 | Dried, milled, sieved | SmF flask | 37°C, pH 3.5–5.5, 150 rpm, $10^6$ spores/mL | 3 | Endopectinase (6 U/mL), Pectinlyase (5 U/mL), Exopectinase (4.8 U/mL), Rhamno-galacturonase (33 U/mL) | Martínez Sabajanes et al. (2012) |

*(Continued)*

**TABLE 3.4 (Continued)**
**Pectinolytic Enzyme Production from Food Wastes**

| Food Waste | Microorganism | Pretreatment Method | Fermentation Mode and Vessel Type | Fermentation Conditions | Duration (day) | Achievements | References |
|---|---|---|---|---|---|---|---|
| Fruit residues (apple, lemon peel, grape skin, and tamarind kernel) | A. terreus FP-370 | Dried, milled, sieved | SmF flask | 37°C, pH 3.5–5.5, 150 rpm, $10^6$ spores/mL | 3 | Endopectinase (3 U/mL), Pectinlyase (33 U/mL), Exopectinase (4.8 U/mL), Rhamno-galacturonase (4 U/mL) | Martínez Sabajanes et al. (2012) |
| Tomato pomace | Aspergillus awamori | Dried, milled, sieved | SSF-plate-type SSF bioreactor | 28°C, pH 5 | 5 | Exo-PG (36.2 IU/g ds) | Umsza-Guez et al. (2011) |
| Lemon peel pomace | Aspergillus niger Aa-20 | Dried, ground | SSF column tray bioreactor | 30°C, 70% MC, 194 mL/min AFR, 2–0.7 mm PS | 4 | Pectinase (2.18 U/mL) | Ruiz et al. (2012) |
| Passion fruit waste | Macrocybe titans | Dried, milled | SSF-250 mL flasks | 28°C in complete darkness | 14 | Pectinase (1.72 U/mL) | Zilly et al. (2012) |
| Orange peel | Aspergillus niger URM5162 | Dried, ground | Fixed bed bioreactor | 25°C, $3.10^5$ spores/mL | 7 | Endo-PG (1.18 U/mL), Exo-PG (4.11 U/mL) | Maclel et al. (2013) |

*Source:* With kind permission from Springer Science+Business Media: *Waste and Biomass Valorization,* Enzyme production from food wastes using a biorefinery concept: A review, 5, 2014a, 903–917, Uçkun, K.E. et al.

*Note:* AJDA: Apple juice depectinizing assay, S: substrate, SSF: solid state fermentation, SmF: submerged fermentation, I/S: inoculum to substrate ratio, AFR: air flow rate, DS: dry substrate, MC: moisture content, PS: particle size; ds: dry solid, PG: polygalacturonase, CMCase: carboxymethylcellulase, NR: not reported.

bagasse with a solid concentration of 16% (w/v) provided good microbial growth and the highest pectinase titer (32 U/mL) was obtained using SSF without adding extra nutrients.

Aeration is another important parameter affecting pectinase production. Umsza-Guez et al. (2011) reported that forced aeration has negative effects on exo-PG synthesis, reducing to half of its activity in a multilayer packed bed reactor. Maciel et al. (2013) obtained the maximum endo- and exo-PG activities of 1.18 U/mL and 4.11 U/mL, respectively, using the reactors without aeration. A system without aeration is advantageous since it is easier to implement and is economical.

The pH of the medium can also affect pectinase production. Martínez Sabajanes et al. (2012) investigated the effect of substrates (apple, lemon peel, grape skin, and tamarind kernel) and fungi (*Aspergillus flavipes* FP-500 and *Aspergillus terreus* FP-370) on the production of pectinases. The highest activities were obtained using lemon peel. With both strains, acidic pH values and high carbon source concentration favored exopectinase and endopectinase production, while higher pH values and low carbon source concentration promoted pectin lyase and rhamnogalacturonase production.

In summary, fruit wastes are superior substrates to produce high titers of pectinolytic enzymes using either SSF or SmF. Process parameters including medium pH, temperature, composition, inoculum size, moisture content, particle size and aeration highly depend on the substrate and microbial strain. Statistical experimental designs can be employed to optimize the fermentation conditions by evaluating the effects and interactions of parameters that rule a biochemical system.

There is no industrial scale FW biorefinery facility currently in operation. However, there are some studies reporting the technical advances and engineering challenges of orange and lemon waste biorefineries (Ángel Siles López et al., 2010; Rivas-Cantu et al., 2013). Direct utilization of citrus peel as animal feed is the simplest option, requiring little infrastructure or investment (Ángel Siles López et al., 2010). However, citrus peel contains high-value compounds such as pectin and D-limonene (Lohrasbi et al., 2010). Pectin is frequently used in food processing, while D-limonene is an essential oil used in the cosmetics, foods, and pharmaceutical industries. D-limonene can be extracted using suitable solvents. The biomass left over after limonene extraction, which mainly consists of pectin and lignocellulose, is an excellent source for pectinolytic and lignocellulolytic enzyme production. Moreover, the residual biomass, that is, lignin, can be used as an energy source.

### 3.2.4 PROTEASES

Proteases are one of the most important commercial enzyme groups because of their wide range of use in the food, pharmaceutical, detergent, dairy, and leather industries (Prakasham et al., 2005; Potumarthi et al., 2007; Chutmanop et al., 2008; Gupta et al., 2012). Some fungal strains such as *Aspergillus, Penicillium*, and *Rhizopus* and bacteria of genus *Bacillus* have been reported as active producers of proteases (Chutmanop et al., 2008; Khosravi-Darani et al., 2008; Jamrath et al., 2012). Although protease production from agro-industrial wastes has been studied in detail using both SSF and SmF, the investigations on the utilization of FW has not been comprehensive. The studies reporting protease production from FW are listed in Table 3.5. Khosravi-Darani et al. (2008) used a newly isolated alkalophilic *Bacillus* sp. in SmF of date wastes without any pretreatment. High activity protease production (57,420 APU/mL) was obtained at pH 10, 37°C, and the enzyme was reported to be thermostable, indicating its possible utilization in industrial applications. Afify et al. (2011) investigated the production of proteases from potato waste in a submerged system using *S. cerevisiae*, and studied the utilization of remaining solid waste as a biofertilizer for plant development. The highest enzyme activity (360 U/mg) was obtained using a fermentation medium containing 15 g/L potato waste, at initial pH 6.0 and 20°C for 72 hours. There are some studies reporting the production of high activity proteases using fishmeal and shrimp wastes. In a study by Gupta et al. (2012), fishmeal from sardine and pink perch were evaluated as the sole carbon and nitrogen sources for alkaline protease production by *Bacillus pumilus* MTCC 7514. The protease (4914 U/mL) was nearly two times higher than that using basal medium (2646 U/mL). Protease production was enhanced to 6966 U/mL and 7047 U/mL when the fermentation was scaled up from flasks to 3.7 and 20 L fermenters, respectively, using fish meal as the sole source (10 g/L). The crude protease was found to have dehairing capacity in leather processing, which is bound to have great environmental benefits in the leather industry. In another study, a powder was prepared from shrimp wastes and tested as a growth substrate for the production of proteases by *P. aeruginosa* MN7 (Jellouli et al., 2008). *P. aeruginosa* MN7 was found to grow and overproduce proteolytic enzymes (15,000 U/mL). Although there are a few reports on protease production from FW, the appreciable protease activities highlight the potential of these wastes.

In addition to its potential utilization in industrial applications, proteases produced from FW can also be used for biorefining different biomasses. Koutinas et al. (2007b) evaluated an oat-based biorefinery for the production of lactic acid as well as other value-added by-products, such as β-glucan and antioxidant-rich oil bodies using *Rhizopus oryzae*. During the process, *Rhizopus oryzae* produced a range of enzymes (glucoamylase, protease, and phosphatase) during the hydrolysis of complex macromolecules in oat. The utilization of waste biomass and *in situ* produced enzyme cocktails in such a biorefining strategy could lead to a significant operating cost reduction as compared to current industrial practices for lactic acid production from pure glucose achieved by bacterial fermentations.

### 3.2.5  LIPASES

After proteases and carbohydrases, lipases (EC 3.1.1.3) are considered the third largest group based on total sales volume (Contesini et al., 2010). They are widely used in the food, detergent, cosmetics, organic synthesis, and pharmaceutical industries. They catalyze the hydrolysis of triacylglycerols to di- and mono-acylglycerols, fatty acids, and glycerol (Alkan et al., 2007; Li et al., 2009; Vaseghi et al., 2013). They are also able to catalyze alcoholysis, acidolysis, aminolysis, esterification, and transesterification under certain conditions (Saxena et al., 2003). Phospholipases are a subclass of lipases that catalyze the hydrolysis of one or more ester and phosphodiester bonds of glycerophospholipids. Their site of action varies on phospholipids,

**TABLE 3.5**

**Protease Production from Food Wastes**

| Food Waste | Microorganism | Pretreatment Method | Fermentation Mode and Vessel Type | Fermentation Conditions | Duration (day) | Achievements | References |
|---|---|---|---|---|---|---|---|
| Date waste | Bacillus sp. 2–5 | Heat treatment and filtration | SmF-125 mL flask | 37°C, pH 10, 125 rpm | 2 | 57,420 APU/mL | Khosravi-Darani et al. (2008) |
| Potato waste | Saccharomyces cerevisiae | NR | SmF-250 mL flask | 28°C | 5 | 360 U/mg | Afify et al. (2011) |
| Fish meal | Bacillus pumilus MTCC 7514 | None | SmF-20L bioreactor | 30°C, pH 7.5 | 2 | 7.05 U/mL | Gupta et al. (2012) |
| Bread waste | Aspergillus oryzae | None | SSF Petri plates | 30°C, MC:1.8 (w/w, db), PS:20 mm, $10^6$ spore/gdS | 6 | 83.2 U/gdS | Melikoglu et al. (2013c) |
| Cuttlefish by-products | Vibrio parahaemolyticus | Heat treatment, pressing, grinding, drying at 80°C o/n, powdering | SmF-250 mL flasks | 37°C, pH 8.7, 200 rpm | 1 | 2487 U/mL | Souissi et al. (2008) |
| Shrimp waste | Pseudomonas aeruginosa MN7 | Heat pretreatment (100°C, 20 min), drying, grinding | SmF-250 mL flasks | 37°C, 200 rpm | <1 | 15,000 U/mL | Jellouli et al. (2008) |

*Source:* With kind permission from Springer Science+Business Media: *Waste and Biomass Valorization*, Enzyme production from food wastes using a biorefinery concept: A review, 5, 2014a, 903–917, Uçkun, K.E. et al.

*Note:* SmF: submerged fermentation, SSF: solid state fermentation, MC: moisture content, PS: particle size, S: substrate, o/n: overnight, NR: Not reported.

which can be used for the modification/production of new phospholipids for some applications in the oil refinery, health, food manufacturing, dairy, and cosmetics industries (Ramrakhiani and Chand, 2011).

Most of the research concentrated on high activity extracellular lipase production using both SmF and SSF via a wide variety of microorganisms including bacteria, fungi, yeast, and Actinomyces (Gupta et al., 2007; Li et al., 2009; Rehman et al., 2011; Vaseghi et al., 2013). Several strains of commercial lipase-producing fungi are quite dominant, including *Rhizopus, Rhizomucor, Aspergillus, Geotrichum, Yarrowia,* and *Penicillium* species (Colen et al., 2006). Recently, the production of lipase was investigated by several researchers using FW as substrates (Alkan et al., 2007) or by supplementing FW as inducer (Dominguez et al., 2010; Papanikolaou et al., 2011). Alkan et al. (2007) investigated the production of lipase from melon waste in SSF using *Bacillus coagulans.* The highest lipase production (78.1 U/g) was achieved after 24 hours of cultivation with 1% olive oil enrichment at 37°C and pH 7.0 by supplementing sodium dodecyl sulphate (Table 3.6). The best results were obtained by supplementing starch and maltose (148.9 and 141.6 U/g, respectively), whereas lower enzyme activities were found in cultures grown on glucose and galactose (approximately 118.8 and 123.6 U/g, respectively). The enzymes were inhibited by $Mn^{2+}$ and $Ni^{2+}$ by 68% and 74%, respectively. By contrast, $Ca^{2+}$ enhanced enzyme production by 5%. In a study by Dominguez et al. (2010), the biodegradation of waste cooking oil and its application as an inducer in lipase production by *Yarrowia lipolytica* CECT 1240 was investigated. The addition of waste cooking oil to the medium led to a significant augmentation in extracellular lipase production by the yeast, compared to oil-free cultures. Papanikolaou et al. (2011) explored the effects of *Aspergillus* and *Penicillium* strains on lipid accumulation and lipase production using waste cooking oil as substrate. In carbon-limited medium, the highest amount of biomass (18 g/L) with a lipid content of 64% was obtained using *Aspergillus* sp. ATHUM 3482, while the highest extracellular lipase activity (645 U/mL) was obtained by *Aspergillus niger* NRRL 363. The studies above indicated the possibility of FW utilization either as substrates or inducers for lipase production. Lipase production can be further improved using mutant or engineered strains.

Lipases are also used for biodiesel production from crude oil and fats (Bajaj et al., 2010) either in free or immobilized form. Lipase production processes from FW can be integrated in a biodiesel biorefining process to decrease the transesterification cost. Phospholipases are used for oil degumming and improving the efficiency of fatty acid yields (Dijkstra, 2010). Although there is no report on phospholipase production using FW, a process for the production of various types of phospholipases from FW can be developed using suitable strains. Further research should be carried out on phospholipase production using FW (Figure 3.2, Pathway B).

## 3.3 CONCLUSIONS

The management of FW poses serious economic and environmental concerns. The publications discussed above indicate that a wide range of industrial enzymes can be produced from FW. The produced enzymes can be used in some industrial

**TABLE 3.6**
**Lipase Production from Food Wastes**

| Food Waste | Microorganism | Pretreatment Method | Fermentation Mode and Vessel Type | Fermentation Conditions | Duration (day) | Achievements | References |
|---|---|---|---|---|---|---|---|
| Banana waste, melon waste, watermelon waste | Bacillus coagulans | None | SSF-Flasks | 37°C, pH 7 | 1 | 148.9 U/g S from melon waste | Alkan et al. (2007) |
| Waste cooking oil | Y. lipolytica CECT 1240 | None | SmF- 5L stirred tank bioreactor with 3L working vol, fb | 30°C, 400 rpm | 6 | 0.93U/mL | Dominguez et al. (2010) |
| Waste cooking olive oil | Aspergillus and Penicillium strains | Filtration | SmF-250 mL flasks | 28°C, pH 6, 200 rpm | 3 | 645 U/ mL | Papanikolaou et al. (2011) |
| Olive oil cake | Y. lipolytica NRLL Y-1095 | Alkaline pretreatment (3% NaOH) 20°C o/n | SSF-150 mL Erlenmeyer fasks | 30°C, pH 7, 55% MC | 4 | 40IU/g S | Moftah et al. (2012) |
| Tri-substrate (wheat bran, wheat rawa, and coconut oil cake) | A. niger MTCC2594 | None | SSF-3*1 kg tray type bioreactor | 30°C, 60% MC | 4 | 745.7 IU/gdS | Edwinoliver et al. (2010) |
| Seafood processing waste | Bacillus altitudinis | Drying (80°C o/n) | SSF-Flasks | 50°C, pH 8, 80% MC | 3 | 2U/gdS (Esterase) | Esakkiraj et al. (2012) |
| Tuna by-products | Rhizopus oryzae | Heat pretreatment (100°C 20 min) and filtration | SmF- 1L flasks | 30°C, pH 6, 150 rpm | 3 | 23.5 IU/mL | Sellami et al. (2013) |
| Wheat bran with 2% olive oil | Aspergillus flavus | None | SSF-Flasks | 29°C, pH 7, 65% MC | 4 | 121.4 U/gdS | Toscano et al. (2013) |
| Wheat bran with 2% olive oil | Aspergillus niger J1 | None | SmF- 500 mL flasks | 30°C, pH 6, 100 rpm | 8 | 1.46 U/mL | Falony et al. (2006) |
| Wheat bran with 2% olive oil | Aspergillus niger J1 | None | SSF-flasks | 30°C, pH 6, 65% MC | 7 | 1.46 U/mL | Falony et al. (2006) |

*Source:* With kind permission from Springer Science+Business Media: *Waste and Biomass Valorization*, Enzyme production from food wastes using a biorefinery concept: A review, 5, 2014a, 903–917, Uçkun, K.E. et al.

*Note:* S: substrate, ds: dry substrate, SSF: solid state fermentation, SmF: submerged fermentation, fb: fed-batch, MC: moisture content, o/n: overnight.

applications. Moreover, these enzyme production processes can be consolidated with other value-added product development processes to create FW biorefineries.

Thus far, all developed biorefinery processes for the conversion of FW into ethanol and other value-added products have only been achieved at benchtop and pilot scales. There is no industrial scale FW biorefinery facility currently in operation. Therefore, it is not yet possible to conduct an economic analysis of the proposed biorefinery systems. However, considering the cost of defined medium preparation in current commercial enzyme processes, the utilization of low or no cost waste biomass for biorefining could lead to significant reductions in operating costs. The difficulties and costs associated with the collection/transportation of FW should therefore also be taken into account. Optimization and scale-up studies need to be carried out in order to exploit these novel FW strategies in large-scale applications.

# 4 Enhanced Glucoamylase Production by *Aspergillus awamori* Using Solid State Fermentation

## 4.1 INTRODUCTION

Due to its chemical complexity, high moisture content, easy degradation, and nutrient-rich composition, food waste should be treated as a useful resource for higher-value products, such as fuels and chemicals through fermentation. Recently, there has been a growing interest in biochemical production from food waste (Han and Shin, 2004; Wang et al., 2005a; Sakai and Ezaki, 2006; Yang et al., 2006; Ohkouchi and Inoue, 2007; Koike et al., 2009; Zhang et al., 2010, 2013b). Starch is an important biopolymer in foods, as such it is a significant part of kitchen waste (Arooj et al., 2008). Hence, the saccharification of food waste is the most important step for its bioconversion into value-added products. For this, commercial enzymes, particularly glucoamylases, are often used to promote the bioconversion of polymers to bioproducts. To produce lactic acid from food waste, Sakai et al. (2004b) used glucoamylase to saccharify the production medium. In other studies, commercial glucoamylase, α-amylase, and cellulase solutions were used to saccharify kitchen wastes for ethanol production (Kim et al., 2008c; Uncu and Cekmecelioglu, 2011; Yan et al., 2012a). If the enzymes were produced *in situ* without downstream treatments and integrated with the biochemical production, the cost of the process could be lower (Merino and Cherry, 2007; Wang et al., 2010c). Moreover, transportation costs and enzyme inactivation during storage could be avoided. If the crude enzyme activity is high, it would be feasible and economical for it to be used directly without any recovery process.

The microorganisms reported to be active producers of amylolytic enzymes are *Aspergillus awamori, Aspergillus foetidus, Aspergillus niger, Aspergillus oryzae, Aspergillus terreus, Mucor rouxians, Mucor javanicus, Neurospora crassa, Rhizopus delmar,* and *Rhizopus oryzae* (Norouzian et al., 2006). Although glucoamylases have been produced by submerged fermentation traditionally, solid state fermentation (SSF) processes have been increasingly applied for the production of this enzyme in recent years (Ellaiah et al., 2002). SSF has advantages over submerged fermentation in that it is simpler, requires less capital investment, has superior productivity, lower energy requirement, requires simpler fermentation media, does not require rigorous control of fermentation parameters, uses less water, produces less wastewater, allows for the easy control of bacterial contamination, and has lower downstream processing costs (Ellaiah et al., 2002; Anto et al., 2006; Melikoglu et al., 2013c).

In order to attain high enzyme activity, a number of factors need to be optimized. Statistical methods for optimization are gaining growing interest and application as they have proved to be cost and time saving. Recently, several statistical experimental design methods were employed for optimizing enzyme production. Among the optimization methods used, central composite design using response surface methodology (RSM) is a method suitable for identifying the effects of individual variables, and for seeking the optimal conditions for a multivariable system efficiently. This approach reduces the number of experiments, improves statistical interpretation possibilities, and reveals possible interactions among parameters. To develop a viable process, it is important to determine the most appropriate substrate and to optimize the fermentation conditions.

Therefore, in this chapter, the effects of food wastes were evaluated to produce glucoamylase using SSF. The objective was therefore to find out the best substrate for glucoamylase production. Then, the fermentation conditions for high-activity glucoamylase production were optimized by RSM. For this, experiments were carried out under different operational conditions defined by four independent variables (initial moisture content, inoculum loading, pH, and incubation time). The objective was also to determine which model best fits the data, and to determine the role of each variable, their interactions, and provide a statistical analysis in order to predict glucoamylase activities.

## 4.2 MATERIALS AND METHODS

### 4.2.1 MATERIALS

*Aspergillus awamori* was used to produce glucoamylase in solid state fermentation for food waste hydrolysis. It was originally obtained from ABM Chemicals Ltd. (Cheshire, England). Its storage and preparation were conducted according to the procedures explained elsewhere (Wang et al., 2007). The cake waste used in this study was collected from local catering. The cake waste were ground, sieved, and then stored at −20°C pending further experiments. The mixed food waste (MFW) used in this study was collected from a local cafeteria. Potatoes, fruits, and vegetables were obtained from a local supermarket. These were discarded from the packaging line because of a lower quality. The food wastes were homogenized in a blender and stored at −20°C pending use in experiments.

### 4.2.2 METHODS

#### 4.2.2.1 Effect of Particle Size on Solid State Fermentation

To determine the effect of particle size, the substrate was sieved through mesh number 5, 10, 16, and 230 corresponding to size cut-off of 0.6, 1.18, 2, and 4 mm, respectively (Endecotts Ltd., UK). After sieving, the moisture content was adjusted to 70%

(wb) and the SSF was carried out with an inoculum loading of $10^6$/g substrate at neutral initial pH and 30°C for 4 days.

### 4.2.2.2 Experimental Design for Enzyme Production

A $2^4$ full factorial design was used in the optimization of glucoamylase production from cake waste. Initial pH ($X_1$), moisture content ($X_2$, %, w/w), inoculum loading ($X_3$, inoculum/g substrate), and time ($X_4$, day) were chosen as independent input variables. The glucoamylase (units/gram dry solid or U/gds) activities were used as dependent output variables. A total of 30 experiments that included 16 cube points (runs 1–16), 8 star points (runs 17–24), and 6 replicas of the central point (runs 25–30) were performed to fit a second-order polynomial model. The experimental range and the levels of the variables are defined and presented in Table 4.1. The ranges of variables used in this work were selected based on literature (Pandey, 1991; Ellaiah et al., 2002; Wang et al., 2009a; Melikoglu et al., 2013d).

### 4.2.2.3 Solid State Fermentation and Enzyme Extraction

Substrates were moistened with the calculated amount of 0.1 M phosphate and citrate buffer solutions in 500 mL Erlenmeyer flasks depending on the targeted initial pHs. After sterilization by autoclaving (120°C for 20 min), the flasks were cooled down, inoculated with inoculum to obtain a certain spores concentration, and the contents were mixed thoroughly with a sterile spatula. Then, the content was distributed into Petri dishes and incubated at 30°C under stationary conditions (Figure 4.1). Petri dishes, in duplicate, were withdrawn at regular time intervals, and the content was extracted with 60 mL of distilled sterile water. This was then centrifuged at 6000 rpm for 10 min and cell-free supernatant was used for assaying the glucoamylase activity.

### 4.2.2.4 Glucoamylase Assay

The activity of glucoamylase was determined at 55°C at neutral pH using 2% (w/v) soluble starch (Sigma) as substrate. The glucose concentration was determined with

---

**TABLE 4.1**

**The Experimental Range and Levels of the Variables in the Central Composite Design**

| Variable | Low Axial ($-\alpha$) | Low Factorial ($-1$) | Center (0) | High Factorial ($+1$) | High Axial ($+\alpha$) |
|---|---|---|---|---|---|
| pH | 5 | 6 | 7 | 8 | 9 |
| Moisture content, (%) | 50 | 60 | 70 | 80 | 90 |
| Inoculum loading (per gs) | $10^3$ | $10^5$ | $5.5 \times 10^5$ | $10^6$ | $1.1 \times 10^6$ |
| Time (days) | 3 | 4 | 5 | 6 | 7 |

*Source:* Uçkun, K.E., A.P. Trzcinski, and Y. Liu. 2014b. *Biofuel Research J.* 3:98–105, with permission from Biofuel Research Journal.

**FIGURE 4.1**   Solid state fermentation for enzyme production. Top: Different food wastes after fungal inoculation on day 0. Bottom: Fungal micelles on cake wastes on day 4.

Optium Xceed blood glucose monitor (Abbott Diabetes Care, Oxon, UK) (Bahcegul et al., 2011). One unit (1 U) of glucoamylase activity was defined as the amount of enzyme releasing 1 micromole of glucose equivalent per minute under the assay conditions.

### 4.2.2.5   Statistical Analysis

The data obtained from the central composite design experiments were analyzed using Design Expert (Stat-Ease Inc., Minneapolis, MN) (Version 8.0.7.1) software, and response surface curves, corresponding contour plots, regression coefficients, and F-values were obtained. Analysis of variance (ANOVA) was applied for the response function. The actions and interactions of the variables were estimated by the following second-order quadratic equation:

$$Y = b_0 + \sum b_i X_i + \sum b_{ij} X_{ij} + \sum b_i^2 X_i^2 \qquad (4.1)$$

where $Y$ is the predicted response for glucoamylase activity (U/gds); $b_0$ is the intercept; $b_i$ is the coefficient for linear direct effect; $b_{ij}$ is the coefficient for interaction effect; $b_i^2$ is the coefficient for quadratic effect (a positive or negative significant value implies possible interaction between the medium constituents); and $X_i$ and $X_{ij}$ are the independent variables. The quality of fit to the second-order equation was expressed by the coefficient of determination ($R^2$) and its statistical significance was determined by the F-test. Variables with probability below 95% ($P > 0.05$) were regarded as not significant for the final model. Three-dimensional surface plots were drawn to illustrate the main and interactive effects of the independent variables on the dependent variables. The influence of experimental error on the central composite design was assessed with six replications at the central point of the experimental domain. Experiments were carried out in triplicate. Results were presented as the average of three independent trials. To maximize the enzyme activity, numerical optimization was used for determination of the optimal levels of the four variables.

### 4.2.2.6 Model Validation

One set of experiments was performed to validate the model. Enzymatic hydrolysis were conducted using an initial pH of 7.9, moisture content of 69.6%, inoculum loading of $5.2 \times 10^5$/gs and incubation time of 6 days to obtain the highest glucoamylase activity: All experiments were performed in triplicate and standard deviations were calculated from the mean of the triplicate analyses.

### 4.2.2.7 Analytical Methods

Moisture and ash contents were determined according to analytical gravimetric methods (AOAC, 2001). Crude protein content was determined using HR Test 'N Tube TN kit (HACH, United States) and calculated according to the Kjeldahl method with a conversion factor of 6.25. Starch content was determined using Megazyme's TN kit (Bray, Ireland). The lipid content was determined by hexane/isopropanol (3:2) method (Hara and Radin, 1978). The glucose concentration was determined with Optium Xceed blood glucose monitor (Abbott Diabetes Care, Oxon, UK) (Bahcegul et al., 2011). Reducing sugars were quantified to monitor the saccharification of food waste according to the dinitrosalicylic acid (DNSA) method using glucose as a standard (Miller, 1959).

## 4.3 RESULTS AND DISCUSSION

### 4.3.1 Effects of Substrate on Glucoamylase Production

In order to understand the effects of different substrates, the wastes were characterized in Table 4.2. The composition of each waste type was different from each other's. Bread has the highest starch content (71.6%) followed by potato (47.6%), cake (45.8%), and savory (45.7%). The reducing sugar content of cake (16.8%), fruit (11.7%), and potato (1.2%) were higher than that of bread (1.5%).

The influence of food wastes such as bread, cake, savory, vegetable, fruit, potato, and mixed type food waste (MFW) from a cafeteria on glucoamylase production by *Aspergillus awamori* was tested for 10 days (Figure 4.2). The incubation time is governed by characteristics of the culture, and is based on growth rate and enzyme production. Maximum glucoamylase production normally occurs after 2–5 days of incubation as reported by other researchers working with solid state cultures involving bacteria and fungi (Soni et al., 2003; Melikoglu et al., 2013d). The fungus used in the present study colonized well on the waste materials, and exhibited a good growth on the surface after 24 hours. The high reducing sugars content in cake, fruit, and potato wastes triggered glucoamylase production so it was higher than savory and MFW on day 1. The growth and enzyme yields improved gradually, and the maximum activity of glucoamylase was obtained using cake wastes on the fourth day of fermentation (Figure 4.2). It was followed by bread, potato, and fruit wastes. The protein content of cake wastes (14.1%) was also higher than that of bread (8.6%). This

**TABLE 4.2**

**Composition of Different Food Wastes**

| FW | Moisture (%) | TS (%) | VS/TS (%) | Starch (%), db | RS (%), db | Protein (%), db | Lipid (%), db | Ash (%), db |
|---|---|---|---|---|---|---|---|---|
| Bread waste | 34.4±0.2 | 65.6±0.2 | 96.7±0.0 | 71.6±0.5 | 0.5±0.1 | 8.6±2.1 | 3.9±2.6 | 3.2±0.0 |
| Cake waste | 29.9±1.9 | 70.1±1.9 | 96.0±0.3 | 45.8±3.0 | 16.8±0.5 | 14.1±0.8 | 16.1±7.5 | 3.9±0.2 |
| Savory | 37.8±0.4 | 62.2±0.4 | 96.6±0.3 | 45.7±2.8 | 0.3±0.0 | 2.3±1.1 | 22.1±0.3 | 3.3±0.4 |
| Discarded fruits | 83.8±2.2 | 16.2±2.2 | 96.6±0.6 | 24.8±4.5 | 11.7±1.5 | 3.5±0.4 | 1.0±0.2 | 3.4±0.6 |
| Discarded potatoes | 82.4±0.7 | 17.6±0.7 | 97.2±0.7 | 47.6±5.5 | 1.2±0.1 | 6.9±2.2 | 0.2±0.0 | 2.7±0.5 |
| Discarded vegetables | 95.2±0.6 | 4.8±0.6 | 85.7±2.0 | 16.4±0.1 | 0.0±0.0 | 0.5±2.2 | 1.5±0.1 | 11.3±1.3 |
| Mixed FW | 80.3±1.1 | 19.7±1.1 | 95.2±0.4 | 19.0±1.3 | 0.7±0.0 | 15.4±2.4 | 19.4±0.1 | 4.7±0.4 |

*Source:*  Uçkun, K.E., A.P. Trzcinski, and Y. Liu. 2014b. *Biofuel Research J.* 3:98–105, with permission from Biofuel Research Journal.

*Note:*  Total solid, starch, reducing sugar (RS) lipid, protein and ash contents are given in wt% on the basis of dry weight (db). Volatile solid (VS) contents are given as the %VS ratio on total solid basis.

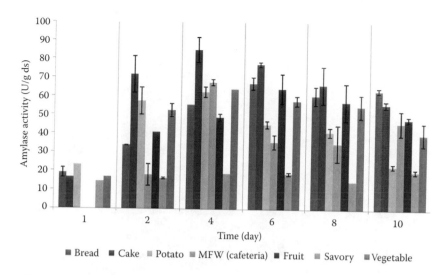

FIGURE 4.2   The effect of substrates on glucoamylase production using a moisture content of 70% (wb), inoculum loading of 10⁶/g substrate at neutral initial pH and 30°C. (From Uçkun, K.E., A.P. Trzcinski, and Y. Liu. 2014b. *Biofuel Research J.* 3:98–105, with permission from Biofuel Research Journal.)

resulted in a better fungal growth and higher glucoamylase activity due to a more balanced composition. As the highest glucoamylase activity was obtained using cake wastes, the following sets of experiments were conducted with that substrate. Liquid state fermentations (also known as submerged fermentation) tests were also carried out (Figure 4.3), but were not as effective as SSF to produce gluco amylase. Bread and cake wastes were equally suitable to produce approximately 40 U/g ds after 4 days (Figure 4.4).

## 4.3.2   EFFECTS OF TEMPERATURE

Many factors affect the enzymatic hydrolysis including temperature, enzyme dose, substrate concentration, and duration. The effect of reaction temperatures on the hydrolysis of domestic FW (10% w/v) using *in situ* produced GA was evaluated in the temperature range of 50°C and 90°C (Figure 4.5). During the first 6 hours, the glucose production was the highest at 70°C (6.59 g/L), and then it slowed down.

FIGURE 4.3   Liquid state submerged fermentation of various substrates.

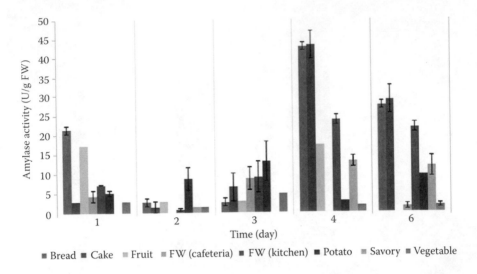

**FIGURE 4.4** The effect of substrates on glucoamylase production during submerged fermentation.

After 6 hours, the glucose production at 50°C and 60°C became higher than that at 70°C. This might be because of enzyme denaturation at temperatures higher than 60°C. These findings are similar to the results reported in the literature. Melikoglu et al. (2013a) evaluated the kinetics of GA production using the same microorganism, and found that the maximum enzyme activity was achieved at 60°C, and then started to decline at higher temperatures due to thermal deactivation of the enzyme. The highest glucose concentration of 10.4 g/L corresponding to a saccharification degree of 97.9% was obtained at 60°C after 24 hours. Hence, the following tests were conducted at 60°C for 24 hours.

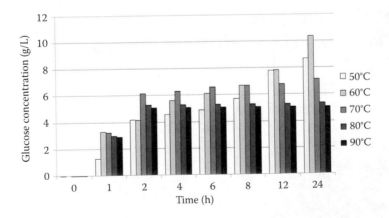

**FIGURE 4.5** The effect of temperature on glucose formation during the hydrolysis of domestic FW with the produced GA solution. Data points show the averages from duplicate analyses for which the standard deviation was <1%.

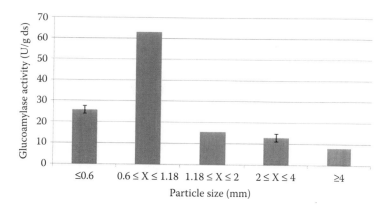

**FIGURE 4.6** The effect of cake particle size on glucoamylase production using a moisture content of 70% (wb), inoculum loading of 10%/g substrate at neutral initial pH and 30°C for 4 days. (From Uçkun, K.E., A.P. Trzcinski, and Y. Liu. 2014b. *Biofuel Research J.* 3:98–105, with permission from Biofuel Research Journal.)

### 4.3.3 EFFECTS OF PARTICLE SIZE

The utilization of the substrate during solid state fermentations by the fungi is not only influenced by its nutritional quality, but also by the particle size of the solid substrate (Schmidt and Furlong, 2012). Results shown in Figure 4.6 validated the fact that particle size has a direct effect on glucoamylase production during solid state fermentation. The highest glucoamylase activity, 63.06 U/gds, was obtained with a particle size of $0.6 \leq X \leq 1.18$ mm. In solid state fermentations, smaller particle size provides larger contact area. However, reduction in particle size increases the packing density, which causes a reduction in the void space between the particles, which results in reduction in microbial growth and enzyme production (Ruiz et al., 2012). Therefore, there is an optimum for particle size. In this study, the size range $0.6 \leq X \leq 1.18$ was chosen in the following set of experiments.

### 4.3.4 EFFECTS OF SUBSTRATE LOADING

To study the effect of FW concentration on the enzymatic hydrolysis, GA treatment was conducted at 10%, 20%, 30%, 40%, and 50% (w/v) FW loadings with 2 U/g FW enzyme loading for 24 hours. The waste concentrations higher than 50% were not studied due to high viscosity of the suspension, which certainly would inhibit enzyme activity. The glucose concentration increased dramatically with an increase in substrate loading, while the hydrolysis continued until the 24 hours (Figure 4.7). The hydrolysis rate was sluggish at waste loadings of 30%, 40%, and 50% during the first 4 hours, and increased dramatically afterward. This might be related to elongated gelatinization of starch. The maximum glucose concentrations of $9.3 \pm 0.9$, $14.8 \pm 0.77$, $19.7 \pm 0.77$, $39.1 \pm 2.93$, and $52.3 \pm 2.97$ g/L were obtained at the respective FW loadings of 10%, 20%, 30%, 40%, and 50% (w/v). Moreover, the saccharification degree reached 99.8% at 50% FW loading. In these experiments,

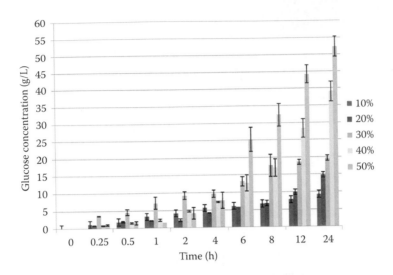

**FIGURE 4.7** The effect of substrate loading on saccharification. The food waste loadings studied were 10%, 20%, 30%, 40%, and 50% (w/v) using glucoamylase (GA) loading of 2 U/g substrate at 60°C for 24 hours.

no substrate inhibition was observed, that is, no dilution of FW was required in the loading range studied, which helps to reduce the generation of wastewater. It should be noted that even after 12 hours, the saccharification degree was found to be 55%–88%, depending on the waste concentration.

As maximizing the production of glucose is the main target, saccharification at higher enzyme loadings should be investigated for increasing glucose concentration, while shortening the hydrolysis time. For this purpose, experiments were carried out at two different GA loadings of 5 and 10 U/g FW, respectively. The activity level of the *in situ* produced enzyme extract was not high enough for treating the suspensions of 50% (w/v) FW; hence, the experiments were conducted in a waste loading range of 10%–40%. Figure 4.8 shows the glucose concentrations obtained using 5 U/g FW and 10%–40% of waste loadings. Almost complete saccharification was achieved within 12 h at 10% and 20% FW loadings, while 24 h were needed at the waste loading of 30% and 40% (Figure 4.8). The hydrolysis rates of 30% and 40% waste suspensions were lower until 2 hours, possibly due to extended gelatinization. Afterward, it increased quickly. The gelatinization process at 5 U/g FW was found to be much shorter than at 2 U/g FW, showing the advantage of using higher enzyme dosages.

The tests using 2 U/g FW and 5 U/g FW were conducted using an extracted enzyme cocktail. The experimental sets using 5 U/g FW for 50% FW loading and the experiments using 10 U/g FW were not possible due to the dilution of enzyme by the extraction. Therefore, these experiments were conducted using crude enzyme cake instead of extracted enzyme solution as it has lower enzyme activity due to the dilution with water. The crude enzyme cake contains the enzymes, substrate (cake waste) residue, and the fungal biomass, and was directly

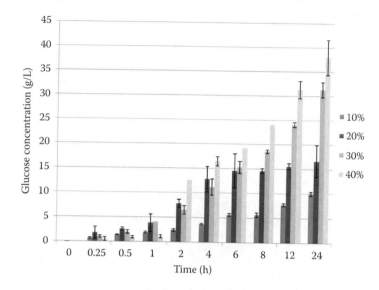

**FIGURE 4.8**   The effect of substrate loading on saccharification. The FW loadings studied were 10%, 20%, 30%, and 40% (w/v) using GA loading of 5 U/g substrate at 60°C for 24 hours. Data points show the averages from duplicate analyses.

added to the FW suspension, without further extraction. The glucose production with 10 U GA/g FW was conducted at waste loadings of 10%, 20%, 30%, and 50% for 24 hours. It was found that glucose production improved significantly, and reached $28.9 \pm 1.44$, $51.7 \pm 1.8$, $86.9 \pm 1.8$ g/L and $115 \pm 7$ g/L at waste loadings of 10%, 20%, 30%, and 50%, respectively (Figure 4.9). Such significant improvement in glucose production is related to a greater starch content of the suspension, which results from the crude enzyme cake.

### 4.3.5   Volume Reduction of Food Waste after Enzymatic Hydrolysis

In this part of the study, the effect of *in situ* produced enzyme solution on the solubilization of FW was investigated. Total suspended solids (TSS) and volatile suspended solids (VSS) were determined at the end of the 24-hours enzymatic hydrolysis. Initial FW suspension contained $49.35 \pm 1.75$ g/L TSS with a $49.08 \pm 1.68$ g/L VSS (Table 4.3). After hydrolysis, 51.1%–62.4% of FW was solubilized. The *in situ* produced enzyme solution significantly improves the hydrolysis of the starch polymer, and also helps to reduce the volume of FW.

### 4.3.6   Utilization of Crude Enzymes for Food Waste Saccharification

FW contains some other carbohydrates like cellulose and hemicellulose apart from starch. Therefore, the addition of cellulases and hemicellulases might further improve the final glucose concentration. For this reason, a fungus (*Trichoderma reesei*) was used to produce crude cellulase enzymes.

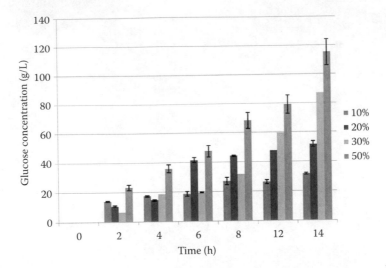

**FIGURE 4.9** The effect of substrate loading on saccharification. The FW loadings studied were 10%, 20%, 30%, and 50% (w/v) using GA loading of 10 U/g substrate at 60°C for 24 hours. Data points show the averages from duplicate analyses.

Various agricultural and kitchen waste residue were assessed for their ability to support the production of cellulase by *Trichoderma reesei* in solid state fermentation. FW such as banana peel, soybean flour, potato peels, oatmeal, and orange waste were used as sole substrate to produce cellulases as the highest cellulase activities were reported using these substrates in the literature. The substrates simply moistened with water (to a 70% final moisture content) were found to be well suited for

**TABLE 4.3**
**Effect of Enzymatic Hydrolysis on VSS and TSS Contents and VSS Reduction**

| Conditions | VSS (g/L) | TSS (g/L) | VSS Reduction (%) |
|---|---|---|---|
| FW (no hydrolysis) | 49.08 ± 1.68 | 49.35 ± 1.75 | – |
| 10% FW suspension with 2 U/g FW GA | 24.00 ± 0.60 | 24.33 ± 0.58 | 51.10 ± 1.22 |
| 10% FW suspension with 5 U/g FW GA | 21.48 ± 3.33 | 22.08 ± 3.58 | 56.24 ± 6.78 |
| 10% FW suspension with 10 U/g FW GA | 20.10 ± 5.70 | 22.78 ± 3.33 | 59.04 ± 11.62 |
| 20% FW suspension with 2 U/g FW GA | 19.30 ± 0.80 | 19.63 ± 0.93 | 60.67 ± 1.63 |
| 20% FW suspension with 5 U/g FW GA | 18.48 ± 0.43 | 26.10 ± 2.85 | 62.35 ± 0.87 |

*Note:* Data points show the averages from duplicate analyses.

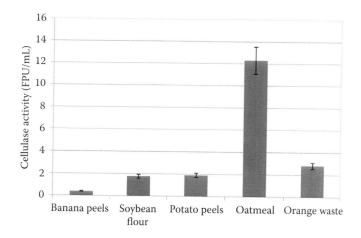

**FIGURE 4.10**    Effect of substrates on cellulase production using *T. reesei* in SSF at 25°C for 6 days. Data points show the averages from duplicate analyses.

fungal growth, producing good amounts of cellulases after 96 h without the supplementation of nutrients. The highest cellulase activity (12.2 FPU/mL) was obtained using oatmeal (Figure 4.10).

The effect of crude enzymes rich in GA and cellulase were evaluated. GA-rich fungal enzyme cocktail resulted in 115 g/L glucose after 24 h hydrolysis, while only 36.5 g/L glucose can be achieved using cellulase-rich enzyme cocktail (Figure 4.11). Even though the enzymatic hydrolysis using GA-rich fungal enzymes cocktail resulted in higher glucose production compared to cellulase, the hydrolysis of complex FW was improved by the co-utilization of both enzyme cocktails together. Using 7 U/g FW GA and 1 FPU/g FW cellulase, 140.1 g/L glucose was produced (Figure 4.11).

### 4.3.7    OPTIMIZATION OF GLUCOAMYLASE PRODUCTION BY RESPONSE SURFACE METHODOLOGY

To determine the optimum pH, moisture content, inoculum loading, and time that maximize glucoamylase activity, 30 experiments were designed using CCD. The experimental conditions and the responses are presented in Table 4.4. A quadratic model was chosen from several models and fitted to the results. The regression equation obtained after the analysis of variance (ANOVA) represented the level of enzyme activity as a function of initial pH, moisture content, inoculum loading, and time.

On the basis of their p-value, $R^2$, SD, and predicted sum of square values, the adequacy of the quadratic regression model was found to be significant for glucoamylase production. The statistical significance of the ratio of mean square variation due to regression and mean square residual error was tested using the ANOVA. The associated p-value is used to estimate whether F is large enough to indicate statistical significance. If the p-value is lower than 0.05, it indicates that the model is statistically significant. The ANOVA result for the glucoamylase production system shows a F-value of 21.96

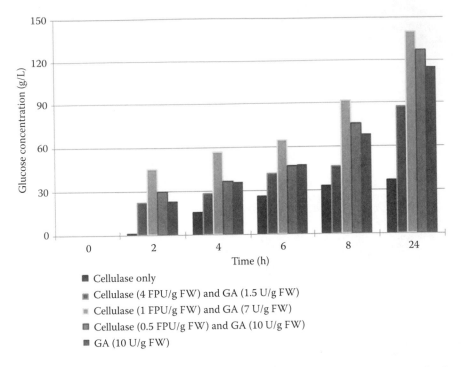

**FIGURE 4.11** Effect of GA- and cellulase-rich enzyme cocktails on glucose production using FW loading of 50%, at 60°C for 24 hours. Data points show the averages from duplicate analyses.

indicating that the model is significant (Table 4.5). There is only a 0.01% chance that a "Model F-Value" this large could occur due to noise. The p-values less than 0.05 indicated that model terms ($X_1$, $X_4$, $X_{14}$, $X_{23}$, $X_{11}$, $X_{22}$, $X_{22}$, $X_{33}$, and $X_{44}$) are significant, and the larger p-values (>0.05) of the regressions $X_2$, $X_3$, $X_{12}$, $X_{13}$, $X_{24}$, and $X_{34}$ suggested their insignificance in the model. The coefficient of determination ($R^2$) for the enzyme activity was calculated as 0.9565, showing that the fitted model could explain 95.65% of variability in the response. Moreover, the high value of $R^2$ indicates that the quadratic equation is able to represent the system under the given experimental domain. An adequate precision of 12.74 for the enzyme activity was recorded. A value greater than 4 is desirable in support of the fitness of the model (Muthukumar et al., 2003). The adjusted $R^2$ corrects the $R^2$ value for the sample size and the number of terms used in the selected model. If there are many terms in the model and the sample size is not large enough, the adjusted $R^2$ may be clearly smaller than $R^2$. The p-value was used to determine the significance of related coefficients. If the p-value is lower than 0.05, the model and the corresponding coefficient are statistically significant (Khuri and Cornell, 1987). The coefficient of variation (CV) indicates the degree of precision with which the treatments are compared. Usually, the higher the CV value, the lower is the reliability of the experiment. In this experiment, a CV value of 22.81 indicates an adjusted greater reliability of the experiments performed. Table 4.5 also shows a term for residual error, which measures the amount of variation in the response data left unexplained by the

## TABLE 4.4
## Central Composite Design with Observed and Predicted Responses of Glucoamylase Activities

| Run | $X_1{}^a$ | $X_2{}^b$ | $X_3{}^c$ | $X_4{}^d$ | Experimental | Predicted |
|---|---|---|---|---|---|---|
| | Actual (Coded) | Actual (Coded) | Actual (Coded) | Actual (Coded) | | |
| 1 | 6 (−1) | 60 (−1) | 100,000 (−1) | 4 (−) | 13.73 | 14.56 |
| 2 | 6 (−1) | 60 (−1) | 100,000 (−1) | 6 (+) | 10.34 | 18.73 |
| 3 | 6 (−1) | 60 (−1) | 1,000,000 (+1) | 4 (−) | 4.2 | −4.59 |
| 4 | 6 (−1) | 60 (−1) | 1,000,000 (+1) | 6 (+) | 2.06 | 12.71 |
| 5 | 8 (+1) | 60 (−1) | 100,000 (−1) | 4 (−) | 36.18 | 35.15 |
| 6 | 8 (+1) | 60 (−1) | 100,000 (−1) | 6 (+) | 92.57 | 84.06 |
| 7 | 8 (+1) | 60 (−1) | 1,000,000 (+1) | 4 (−) | 4.14 | 3.73 |
| 8 | 8 (+1) | 60 (−1) | 1,000,000 (+1) | 6 (+) | 67.34 | 65.81 |
| 9 | 6 (−1) | 80 (+1) | 100,000 (−1) | 4 (−) | 5.76 | 5.66 |
| 10 | 6 (−1) | 80 (+1) | 100,000 (−1) | 6 (+) | 5.26 | 7.42 |
| 11 | 6 (−1) | 80 (+1) | 1,000,000 (+1) | 4 (−) | 7.34 | 17.60 |
| 12 | 6 (−1) | 80 (+1) | 1,000,000 (+1) | 6 (+) | 33.14 | 32.53 |
| 13 | 8 (+1) | 80 (+1) | 100,000 (−1) | 4 (−) | 26.62 | 17.72 |
| 14 | 8 (+1) | 80 (+1) | 100,000 (−1) | 6 (+) | 57.08 | 64.24 |
| 15 | 8 (+1) | 80 (+1) | 1,000,000 (+1) | 4 (−) | 27.46 | 17.43 |
| 16 | 8 (+1) | 80 (+1) | 1,000,000 (+1) | 6 (+) | 76.24 | 77.13 |
| 17 | 7 (0) | 50 (−α) | 550,000 (0) | 5 (0) | 39.9 | 40.14 |
| 18 | 9 (+α) | 70 (0) | 550,000 (0) | 5 (0) | 51.00 | 62.24 |
| 19 | 7 (0) | 70 (0) | 1000 (−α) | 5 (0) | 53.41 | 53.08 |
| 20 | 7 (0) | 70 (0) | 1,100,000 (+α) | 5 (0) | 6.58 | 15.70 |
| 21 | 7 (0) | 70 (0) | 550,000 (0) | 3 (−α) | 88.78 | 79.54 |
| 22 | 7 (0) | 70 (0) | 550,000 (0) | 7 (+α) | 93.20 | 88.44 |
| 23 | 5 (−α) | 70 (0) | 550,000 (0) | 5 (0) | 8.44 | −2.92 |
| 24 | 7 (0) | 90 (+α) | 550,000 (0) | 5 (0) | 42.88 | 42.52 |
| 25 | 7 (0) | 70 (0) | 550,000 (0) | 5 (0) | 93.67 | 88.44 |
| 26 | 7 (0) | 70 (0) | 550,000 (0) | 5 (0) | 74.86 | 88.44 |
| 27 | 7 (0) | 70 (0) | 550,000 (0) | 5 (0) | 93.67 | 88.44 |
| 28 | 7 (0) | 70 (0) | 550,000 (0) | 5 (0) | 81.38 | 88.44 |
| 29 | 7 (0) | 70 (0) | 550,000 (0) | 5 (0) | 93.67 | 88.44 |
| 30 | 7 (0) | 70 (0) | 550,000 (0) | 5 (0) | 117.86 | 88.44 |

*Source:* Uçkun, K.E., A.P. Trzcinski, and Y. Liu. 2014b. *Biofuel Research J.* 3:98–105, with permission from Biofuel Research Journal.

*Note:* Each row corresponds to a single experiment.

[a] Initial pH.

[b] Moisture content (%, w/w).

[c] Inoculum loading (inoculum/g substrate).

[d] Time (day).

**TABLE 4.5**
**ANOVA for Glucoamylase Production as a Function of Initial pH ($X_1$), Moisture Content ($X_2$), Inoculum Loading ($X_3$), and Time ($X_4$)**

| Source | Sum of Squares | DF | Mean Square | F-Value | p-Value Prob > F |
|---|---|---|---|---|---|
| Model | 31703.19 | 14 | 2264.51 | 21.96 | <0.0001 |
| $X_1$-pH | 6367.44 | 1 | 6367.44 | 61.75 | <0.0001 |
| $X_2$-moisture content | 8.52 | 1 | 8.52 | 0.083 | 0.7780 |
| $X_3$-inoculum loading | 42.37 | 1 | 42.37 | 0.41 | 0.5318 |
| $X_4$-time | 6112.04 | 1 | 6112.04 | 59.28 | <0.0001 |
| $X_{12}$ | 72.25 | 1 | 72.25 | 0.70 | 0.4166 |
| $X_{13}$ | 149.57 | 1 | 149.57 | 1.45 | 0.2484 |
| $X_{14}$ | 2003.91 | 1 | 2003.91 | 19.43 | 0.0006 |
| $X_{23}$ | 969.39 | 1 | 969.39 | 9.40 | 0.0084 |
| $X_{24}$ | 5.66 | 1 | 5.66 | 0.055 | 0.8181 |
| $X_{34}$ | 173.45 | 1 | 173.45 | 1.68 | 0.2156 |
| $X_{11}$ | 5660.75 | 1 | 5660.75 | 54.90 | <0.0001[a] |
| $X_{22}$ | 3636.26 | 1 | 3636.26 | 35.27 | <0.0001[a] |
| $X_{33}$ | 3628.09 | 1 | 3628.09 | 35.19 | <0.0001[a] |
| $X_{44}$ | 2730.15 | 1 | 2730.15 | 26.48 | 0.0001 |
| Residual | 1443.52 | 14 | 103.11 | | |
| Lack of fit | 1104.55 | 9 | 122.73 | 1.81 | 0.2660 |
| Pure error | 338.97 | 5 | 67.79 | | |
| Corrected total | 33146.71 | 28 | | | |

*Source:* Uçkun, K.E., A.P. Trzcinski, and Y. Liu. 2014b. *Biofuel Research J.* 3:98–105, with permission from Biofuel Research Journal.

[a] Significant variable; DF, degree of freedom; determination coefficient ($R^2$), 0.9565; adjusted determination coefficient ($R^2$adj), 0.9129; coefficient of variation (CV), 22.81; adequate precision ratio, 12.74.

model. The analysis shows that the form of the model chosen to explain the relationship between the factors and the response is correct.

Equation 4.2 in terms of actual factors (confidence level above 95%) as determined by design of expert software is given below:

$$GA\ Activity\ (U/gds) = -1,366.16 + 184.69X_1 + 17.38X_2 + 8.02*10^{-6}X_3 + 39.82X_4$$
$$-0.21X_1X_2 - 6.79*10^{-6}X_1X_3 + 11.19X_1X_4 + 1.73*10^{-6}X_2X_3 - 0.21X_1X_2$$
$$-6.79*10^{-6}X_1X_3 + 11.19X_1X_4 + 1.73*10^{-6}X_2X_3 - 0.06X_2X_4 + 7.32*10^{-6}X_3X_4$$
$$-14.7X_1^2 - 0.12X_2^2 - 1.11*10^{-10}X_3^2 - 10.21X_4^2$$

$$(4.2)$$

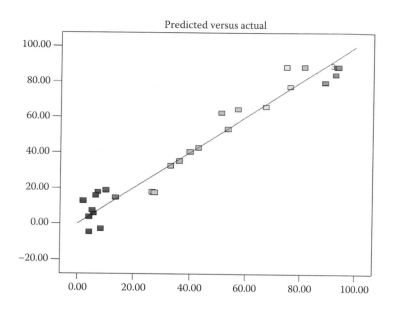

**FIGURE 4.12** The observed (X axis) versus the predicted (Y axis) glucoamylase activities under the experimental conditions. (From Uçkun, K.E., A.P. Trzcinski, and Y. Liu. 2014b. *Biofuel Research J.* 3:98–105, with permission from Biofuel Research Journal.)

where $X_1$, $X_2$, $X_3$, and $X_4$ are independent variables representing the pH, moisture content, inoculum loading, and time, respectively. The negative coefficients for $X_{12}$, $X_{13}$, $X_{24}$, $X_{11}$, $X_{22}$, $X_{33}$, and $X_{44}$ demonstrate the existence of quadratic and linear interaction effects that decrease the response quantity, while the positive coefficients for $X_{14}$, $X_{23}$, and $X_{34}$ expose the existence of quadratic interaction effects that enhance the activity of glucoamylase. Figure 4.12 shows the correlation between the experimental and predicted values of the response. The points close to the line indicate a good fit between the experimental and predicted data.

The optima of the variables for which the responses are maximized are represented by the contour plots (Figure 4.13). The contour plot of the moisture content and pH effect on the activity of glucoamylase illustrates that neutral pHs led to higher enzyme activity using an initial moisture content of around 66%–74% (wb) (Figure 4.13a). The maximum activity of 92.92 U/gds was determined at pH 7.5 using an initial moisture content of 69.6%. Lower initial moisture content provides lower solubility of the nutrients, while higher moisture contents cause decreased porosity and decrease in gas exchange. The moisture content range is consistent with the levels reported in the literature for solid state fermentation of bread waste and wheat flour by *A. awamori* (Wang et al., 2009a; Melikoglu et al., 2013d). Generally, the initial pH for glucoamylase production by *A. awamori* using SSF is adjusted to neutral pHs as the fungus grows well at such pHs. Since the maximum activity of 92.92 U/gds was determined at pH 7.5 using an initial moisture content of 69.6%, these conditions were kept constant in the subsequent studies to find the optimum inoculum loading and incubation time.

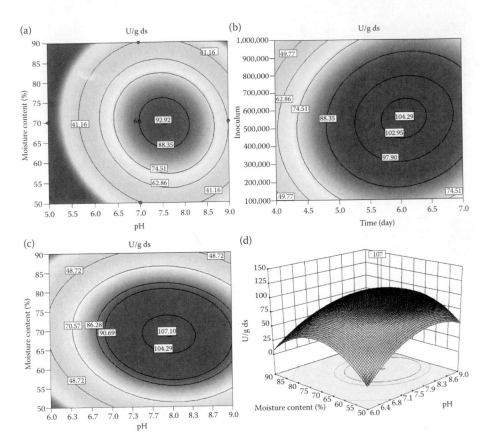

**FIGURE 4.13** Contour plots, described by Equation 4.2, representing the effect of initial pH and moisture content using inoculum loading of $5 \times 10^5$/g substrate for 5 days (a); inoculum loading and incubation time using the initial moisture content of 69.6% and pH of 7.5 (b); inoculum loading and pH using initial moisture content of 69.6% for 6 days (c); initial pH and moisture content using inoculum loading of $5.2 \times 10^5$/g substrate for 6 days (d) on glucoamylase activity from cake waste. (From Uçkun, K.E., A.P. Trzcinski, and Y. Liu. 2014b. *Biofuel Research J.* 3:98–105, with permission from Biofuel Research Journal.)

The glucoamylase production increased using an inoculum loading of $2 \times 10^5$ to $9 \times 10^5$/gs for 5 to 7 days and the maximum glucoamylase activity of 104.29 U/gds was obtained using $5.2 \times 10^5$/gs inoculum on the sixth day of the fermentation (Figure 4.13b). During the fermentation, medium pH, nutrient concentration, temperature, moisture content, and physical structure of the raw material change constantly. All these parameters affect microbial growth and enzyme production. According to Melikoglu et al. (2013d), the growth of *A. awamori* on bread pieces increased exponentially between the third and fifth days, and glucoamylase production reached its maximum level on the sixth day of fermentation. However, as the medium pH was not controlled, the pH is decreasing during this period. Melikoglu et al. (2013d) reported that the pH decreased to 3.8 on the fifth day of the fermentation. This may be one of the major causes of deceleration of the growth and enzyme production after the sixth day of fermentation. Therefore, the effect of initial pH was evaluated using the optimized parameters, and it was predicted that the glucoamylase activity increased from 90.69 U/gds to 107.1 U/gds using initial pH of 7.9 instead of pH 7.0 (Figure 4.13c). The pH reached 4.5 after the fifth day of fermentation when the initial pH was 8 and 9. On the other hand, the pH decreased to 3.5 and 4 when the initial pH was adjusted to 6 and 7, respectively. This explains why the microbial growth and glucoamylase production was enhanced using an initial pH of 7.9.

### 4.3.8 VALIDATION OF THE RESPONSE MODEL

To evaluate the accuracy of the quadratic polynomial model, a verification experiment was conducted under the predicted optimal conditions, and the result was 108.47 U/gds, which is 1.37% higher than the predicted value. This is higher than values reported by Wang et al. (2009a) with the same fungus using wheat flour and similar to those reported by Melikoglu et al. (2013d) on bread pieces. This high degree of accuracy obtained confirms the validity of the model with minor discrepancy due to the slight variation in experimental conditions. The activity obtained is 1.4-fold higher than the yield obtained with cake wastes at the sixth day of fermentation without optimization, suggesting the important role of RSM for rapid screening of important process variables in the optimization studies.

## 4.4 CONCLUSIONS

This chapter demonstrated the feasibility of effective production of glucoamylase with solidstate fermentation using food wastes as the sole nutrient source. Glucoamylase with the highest activity was produced from cake waste using solid state fermentation by *A. awamori*. The optimum conditions for glucoamylase production from cake waste were determined as an initial pH of 7.9, initial moisture content of 69.6%, inoculum loading of $5.2 \times 10^5$ spores/gs, and incubation time of 6 days. Under these conditions, a glucoamylase activity of 108.47 U/gds was obtained. This study showed that cake waste could be an ideal raw material for production of high-activity enzymes through solid state fermentation. Subsequently, a fraction of this high GA activity fungal mash can be used to saccharify FW, and produce a medium of high glucose concentration that is suitable for a wide range of fermentation. Using only 7 U GA/FW and 1 U cellulase/FW, a glucose concentration of 140 g/L can be obtained using this strategy.

# 5 Enhancing the Hydrolysis and Methane Production Potential of Mixed Food Wastes by an Effective Enzymatic Pretreatment

## 5.1 INTRODUCTION

Food waste (FW) is a worldwide issue and in rich countries, it is incinerated with other combustible municipal wastes for volume reduction and recovery of heat and energy, while the residual ash is disposed of in landfills. It should be noted that incineration is not a preferable option of FW management because of to the high moisture content of FW, high operational cost, and generation of hazardous ashes and greenhouse gases (e.g., carbon dioxide) (El-Fadel et al., 1997). On the contrary, FW should be considered a useful resource for producing high-value products (e.g., biofuels and platform chemicals) due to its organic-rich nature.

Anaerobic digestion of FW has been widely studied for biogas generation, which is a viable option for volume reduction and energy recovery from FW (Uçkun et al., 2014a–d). Although FW is readily biodegradable with a volatile solid fraction of up to 90%, the hydrolysis of solid FW into soluble organics has been known as the rate-limiting step of anaerobic digestion (Zhang et al., 2014). As a result, anaerobic digestion of FW has the drawbacks of long solid residence time and low conversion efficiency, indicating that a large anaerobic reactor is required (Quiroga et al., 2014). Therefore, pretreatment methods of FW have been investigated for enhancing the hydrolysis of FW, for example, ultrasonication (Li et al., 2013), microwave (Marin et al., 2010), thermochemical (Cavaleiro et al., 2013), and enzymatic hydrolysis (Moon and Song, 2011). Commercial enzymes including carbohydrases, such as glucoamylase, arabinase, cellulase, β-glucanase, hemicellulase, xylanase, proteases, and lipases have been used to improve the hydrolysis of starch in FW (Moon and Song, 2011). The pretreatment of FW with multiple commercial enzymes appears to be more efficient than with a single commercial enzyme (Kim et al., 2006a; Moon and Song, 2011). However, it should be noted that commercial enzymes are costly (e.g., about $120 for treating 1 ton of FW with glucoamylase and α-amylase at 10 U/g FW) and generally available in single-type form. In order to make the enzymatic hydrolysis of FW more cost-effective, the enzymes should be produced *in situ* from a cheap feedstock without complex and costly downstream separation and purification steps.

Thus far, various kinds of food wastes have been used to produce enzymes including proteases, cellulases, amylases, lipases, and pectinases particularly through solid state fermentation (SSF). Higher enzyme yields can be obtained using SSF as it provides a similar environment to the microorganism's natural environment which provides better conditions for its growth and enzyme production (Thomas et al., 2013). Melikoglu (2008) developed a multi-enzymes solution of glucoamylase and protease during SSF of bread waste using *A. awamori*. This solution was used for the hydrolysis of bread waste and wheat flour. Recently, this concept was also applied for the enzymatic hydrolysis of mixed food waste (MFW) to produce a fermentation medium (Pleissner et al., 2014), which was further used as a nutrient-complete feedstock for the cultivation of microalgae (Pleissner et al., 2013; Lau et al., 2014) and succinic acid production (Sun et al., 2014). Therefore, this chapter aims to (1) produce report the production of a fungal mash rich in glucoamylase with FW as feedstock, and (2) investigate its application for the enzymatic pretreatment of FW to enhance hydrolysis, biomethane production, and waste volume reduction.

## 5.2 MATERIALS AND METHODS

### 5.2.1 CAKE WASTE FOR THE PRODUCTION OF FUNGAL MASH

In Chapter 4, bakery wastes, particularly cake waste, were found to be a good substrate for glucoamylase (GA) production. *A. awamori* was used to produce GA with cake waste collected from a local catering as substrate through SSF. The cake waste was first ground, sieved, and then stored at −20°C for further experiments. MFW used in this study was collected from a local cafeteria, and was homogenized in a blender immediately after collection. The homogenized FW was then stored at −20°C for further use. The composition of cake waste and MFW are presented in Table 5.1. For the purpose of comparison, two commercial enzymes α-amylase and glucoamylase, were also employed. The optimal pH ranges for α-amylase and glucoamylase were 5.0–5.8 and 4.2–4.8, respectively. It appears from Table 5.1 that proteins are not the main component of FW, thus commercial proteases were not tested. Nevertheless, free amino nitrogen concentration in the hydrolyzate was determined, which indicates the presence of proteases in the fungal mash.

### 5.2.2 PRODUCTION OF FUNGAL MASH

Cake waste with a particle size of 1.2–2.0 mm were used as the sole carbon source for producing an enzyme cocktail using SSF in which moisture content was adjusted to 70% (wb) with 0.1 M phosphate buffer (pH 7.9). After sterilization by autoclaving at 120°C for 20 minutes, the flasks were cooled down, and then inoculated with *A. awamori* to obtain a spore concentration of $10^6$/g substrate, and the contents were mixed thoroughly with a sterile spatula. Ten grams of such mixture was distributed into several identical Petri dishes and incubated at 30°C for 6 days under stationary conditions. The GA activity of the fungal mash harvested from two identical Petri

**TABLE 5.1**

**Composition of Food Wastes per Gram of Dry Mass**

|  | Starch (mg) | Reducing Sugar (mg) | Protein (mg) | Lipid (mg) | Ash (mg) |
|---|---|---|---|---|---|
| Cake waste | $458 \pm 30$ | $168 \pm 5$ | $141 \pm 8$ | $161 \pm 7.5$ | $39 \pm 2$ |
| Mixed FW | $461 \pm 32$ | $82 \pm 7$ | $111 \pm 18$ | $153 \pm 21$ | $21 \pm 1$ |

*Source:* Reprinted from *Bioresource Technology*, 183, Uçkun, K.E., A.P. Trzcinski, and Y. Liu., Enhancing the hydrolysis and methane production potential of mixed food wastes by a cost-effective enzymatic pretreatment. 47–52, Copyright 2015, with permission from Elsevier.

dishes was found to be $113.7 \pm 5.2$ U/g dry solids. The fungal mash, that is, the GA-rich fermented solids, were obtained at the end of the fermentation, and it was directly used to hydrolyze MFW without further separation of produced enzymes.

## 5.2.3 HYDROLYSIS OF FOOD WASTE

Blended domestic FW was inoculated with the fungal mash produced at a substrate loading of 50% (w/v) and a GA loading of 10 U/g dry FW. Hydrolysis was performed in duplicate in Duran bottles in a water bath shaker at 60°C and 100 rpm for 24 hours. For the purpose of comparison, similar experiments were also conducted in duplicate with commercial enzymes at 8.6 U/g dry FW for α-amylase and 10 U/g dry FW for GA. Samples taken at different time intervals were centrifuged at 10,000 rpm for 5 minutes before the analyses. The hydrolysis efficiency and solid mass reduction were determined by soluble COD and content of volatile suspended solids after the pretreatments. The detailed experimental procedure is presented in Figure 5.1.

## 5.2.4 ANAEROBIC DIGESTION OF ENZYMATICALLY PRETREATED FOOD WASTE

The inoculum used for the anaerobic digestion was taken from a local full-scale anaerobic digester. After filling up the bottles with the respective amounts of pretreated FW (281 mg TS, 270 mg VS), 36.35 mL inoculum (32.3 g/L TSS, 14.87 g/L VSS), and anaerobic biomedium (30 mL), the headspace was purged with $N_2$ gas at 1 L/min for 3 min, and was then sealed immediately with rubber lids and metal caps to maintain anaerobic conditions (Trzcinski and Stuckey (2012). Biochemical methane potential (BMP) of FW was determined in duplicate on an orbital shaker operated at 35°C and 150 rpm.

## 5.2.5 ANALYTICAL METHODS

Moisture and ash contents of FW were determined by analytical gravimetric methods (AOAC, 2001). Crude protein content was measured using HR Test 'N Tube TN kit (HACH, United States) and calculated according to the Kjeldahl method

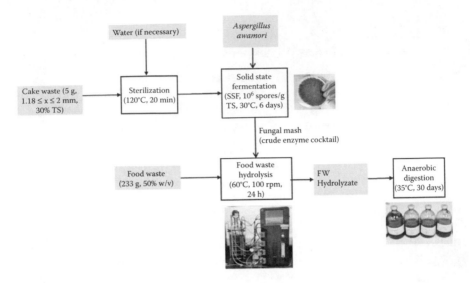

**FIGURE 5.1** Experimental procedure. (Reprinted from *Bioresource Technology*, 183, Uçkun, K.E., A.P. Trzcinski, and Y. Liu., Enhancing the hydrolysis and methane production potential of mixed food wastes by a cost-effective enzymatic pretreatment. 47–52, Copyright 2015, with permission from Elsevier.)

with a conversion factor of 6.25. Starch content was determined using Megazyme's TN kit (Bray, Ireland). The lipid content was determined by hexane/isopropanol (3:2) method (Hara and Radin, 1978). The glucose concentration was determined with Optimum Xceed blood glucose monitor (Abbott Diabetes Care, Oxon, UK) (Bahcegul, 2011). Reducing sugars were quantified to monitor the saccharification of FW according to the dinitrosalicylic acid (DNSA) method using glucose as a standard (Miller, 1959). Free amino nitrogen (FAN) concentration was measured in hydrolyzates using the ninhydrin reaction method (Lie, 1973). Soluble COD and volatile suspended solid reduction were determined using the standard methods (APHA-WPCF, 1998).

Protease activity was estimated through the formation of FAN by hydrolyzing 15 g/L casein solution (Sigma) at 60°C in 200 mM of citrate buffer at pH 4.8. One unit activity (U) was defined as the protease required for the production of 1 g FAN in 1 min. GA activity was determined with 2% (w/v) of soluble starch (Sigma) as substrate at 60°C and pH 4.8. One unit (1 U) of GA activity was defined as the amount of enzyme releasing one micromole glucose equivalent per minute under the assay conditions. All the analytical assays were conducted in triplicate.

The production yield and rate of biogas during anaerobic digestion of FW with and without pretreatment was evaluated by standard BMP tests. The contents of biogas were analyzed by gas chromatography (Agilent 7890A) equipped with a thermal conductivity detector (TCD) and a HayeSep capillary column. The operational temperatures of the injector, detector, and column were set at 100, 150, and 115°C, respectively. Helium at a flow rate of 35 mL/min was used as a carrier gas.

## 5.2.6  DATA ANALYSIS

In this chapter, the modified Gompertz equation was used for comparing the kinetics of methane production from FW with and without pretreatment (Li et al., 2013):

$$B = B_0 \exp\left\{-\exp\left[\frac{R_m e}{B_0}(\lambda - t) + 1\right]\right\} \tag{5.1}$$

where $B_0$ is the estimated ultimate cumulative methane yield or methane production potential (mL/g VS), B is the cumulative methane yield (mL/g VS) at incubation time t (h), e is equal to 2.7183, $R_m$ is the maximum methane production rate (mL/g VS h), and $\lambda$ is the lag phase time (h).

## 5.3  RESULTS AND DISCUSSION

### 5.3.1  PRODUCTION OF GLUCOSE AND FREE AMINO NITROGEN FROM FOOD WASTE PRETREATED WITH FUNGAL MASH

MFW collected from a cafeteria was pretreated, respectively, with fungal mash produced in this study and commercial enzymes (e.g., α-amylase and glucoamylase). It can be seen in Figure 5.2a that the highest glucose concentration of 99.1 ± 7 g/L was obtained after 24 hours in the experiment supplied with the fungal mash, while 77.2 ± 6.9 g/L was reached with commercial enzymes. It should also be noted that the initial glucose production rate with the fungal mash was significantly higher than with commercial enzymes. Given the complex composition of FW, a cocktail of enzymes are required for high-efficiency hydrolysis and saccharification. Cekmecelioglu and Uncu (2013) developed a complex and costly pretreatment procedure using several enzymes, namely α-amylase, glucoamylase, cellulase, and β-glucosidase. The highest glucose concentration achieved was only 64.8 g/Ls achieving 70% conversion after 6 hours of enzymatic hydrolysis of FW. This is lower than the glucose concentration of 86 g/L obtained in this study after 4 hours hydrolysis of mixed FW with fungal mash (Figure 5.2a). Although *A. awamori* is known to be an efficient producer of glucoamylases, it can also produce other hydrolytic enzymes, such as amylases, proteases, cellulases, and xylanases when growing on complex substrates, such as MFW in SSF (Koutinas et al., 2007a; López et al., 2013). For instance, it had been reported that the fermented solids obtained from the SSF of babassu cake with *A. awamori* contained considerable activities of proteases, xylanases, and cellulase besides amylases (López et al., 2013). Table 5.2 shows that about 90%–95% of starch in FW was hydrolyzed by the fungal mash produced by this method. This in turn suggests that this fungal mash contained some other carbohydrases such as

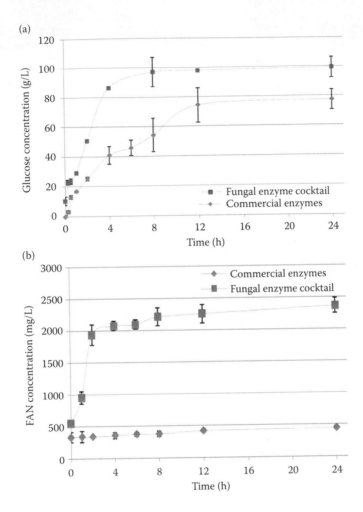

**FIGURE 5.2** Effect of enzymatic pretreatment on glucose (a) and FAN (b) production from FW. Each data point is the average of triplicate measurements from duplicate experiments and the error bars represents the standard deviations. (Reprinted from *Bioresource Technology*, 183, Uçkun, K.E., A.P. Trzcinski, and Y. Liu., Enhancing the hydrolysis and methane production potential of mixed food wastes by a cost-effective enzymatic pretreatment. 47–52, Copyright 2015, with permission from Elsevier. )

α-glucosidases, β-amylases, β-glucanases pullulanases, cellulases, xylanases, and hemicellulases, besides glucoamylase. Compared to commercial enzymes, the fungal mash produced *in situ* offers advantages over extracted enzymes by avoiding the enzyme extraction step, and thereby reducing economical constraint.

The ultimate glucose concentration and the time required for the hydrolysis are related to the moisture and carbohydrate content of the food waste, enzymes, substrate loadings and process parameters. The conditions used in different studies using food wastes were very different from each other (Table 5.3). In a study by

**TABLE 5.2**

**Glucose, FAN, and sCOD Released from the Hydrolysis of Food Waste after 24 h Hydrolysis**

| | Fungal Mash | | Commercial Enzymes | |
|---|---|---|---|---|
| | Concentration (g/L) | Conversion Yield | Concentration (g/L) | Conversion Yield |
| Glucose | 89.1 ± 7.0 | 90%–95%[a] | 77.2 ± 6.9 | 73%–87%[a] |
| FAN | 1.94 ± 0.12 | 72%–80%[b] | 0.10 ± 0.0 | 4%[b] |
| sCOD | 164.7 ± 16.7 | NA | 128.5 ± 2.9 | NA |
| VS reduction | NA | 64% | NA | 52% |

*Source:* Reprinted from *Bioresource Technology*, 183, Uçkun, K.E., A.P. Trzcinski, and Y. Liu., Enhancing the hydrolysis and methane production potential of mixed food wastes by a cost-effective enzymatic pretreatment. 47–52, Copyright 2015, with permission from Elsevier.

*Note:* NA: Not applicable.

[a] On starch basis (all the glucose produced was accounted for by the breakdown of starch).

[b] On protein basis.

Pleissner et al. (2014), a glucose concentration of 143 g/L was obtained after 48 h fermentation at an FW loading of 43.2% (w/v) when solid mashes of *A. awamori* and *A. oryzae* were successively added at an interval of 24 h.

The release of proteins during the hydrolysis of FW by the fungal mash was determined in terms of FAN (Figure 5.2b). It was found that the release of dissolved proteins was negligible during the pretreatment of FW with commercial enzymes as they did not contain any protease. In contrast, the total FAN content in the hydrolyzates obtained during the pretreatment of FW with the fungal mash produced quickly reached 1.94 g/L after 2 hours, and then stabilized at 2.4 g/L after 24 hours. This can be explained by the protease activity detected in the fungal mash (1.37 ± 0.4 U/gds). The total nitrogen analysis revealed that 72%–80% of proteins in FW were solubilized by the fungal mash. Moreover, proteases are known to hydrolyze carbohydrates by breaking down the bindings of proteins (Kim et al., 2006a). Hence, the solubilization of FW was enhanced through the synergistic actions of enzymes present in the fungal mash. It should be noted that high FAN concentration is essential for subsequent fermentation as it provides a balanced nitrogen source for bacterial metabolism and growth. In this study, the bio-available carbon to nitrogen (C/N) ratios were 16.7 and 68.4 in the hydrolyzates obtained from the FW pretreatment with the fungal mash and commercial enzymes, respectively. It has been reported that a feedstock with a C/N ratio greater than 30 is considered deficient in nitrogen for a biological treatment process (Kayhanian and Rich, 1995; Gomez et al., 2005). Therefore, the hydrolyzate obtained from the hydrolysis of FW with the fungal mash is a good biomedium for subsequent biological processes, for example, ethanol fermentation or anaerobic digestion.

**TABLE 5.3**

**Glucose Concentrations and Yields Achieved Using Food Waste**

| Enzymes | $C_{Glucose}$ (g/L) | Carbohydrate Conversion Rate (%) | Duration (h) | References |
|---|---|---|---|---|
| GA, protease, cellulase | 69.8 | 63[a] | 12 | Kim et al. (2011b) |
| GA, cellulase, α-amylase, β-glucosidase | 64.8 | 70 | 6 | Cekmecelioglu and Uncu (2013) |
| GA, cellulase, α-amylase, β-glucanase, xylanase, hemicellulase, arabinase | 79.1 | NR | 8 | Jeong et al. (2012) |
| GA, cellulase, α-amylase, β-glucanase, xylanase, hemicellulase, arabinase | 58.0 | 46[a] | 6 | Moon et al. (2009) |
| GA, α-amylase, β-glucosidase | 65.0 | NR | 24 | Hong and Yoon (2011) |
| GA, α-amylase, protease | 119.2 | 66[a] | 24 | Sun et al. (2014) |
| Fungal mash (*A. awamori* and *A. oryzae*) | 143.0 | 80–90[a] | 48 | Pleissner et al. (2014) |
| GA, α-amylase | 74.3 | 78–82 | 12 | This study |
| Fungal mash (*A. awamori*) | 87.7 | 85–95 | 12 | This study |

*Source:* Reprinted from *Bioresource Technology*, 183, Uçkun, K.E., A.P. Trzcinski, and Y. Liu., Enhancing the hydrolysis and methane production potential of mixed food wastes by a cost-effective enzymatic pretreatment. 47–52, Copyright 2015, with permission from Elsevier.

*Note:* GA: glucoamylase, C: concentration, NR: not reported.

[a] Substrate conversion rate (g glucose/g dry food waste).

## 5.3.2 RELEASE OF SOLUBLE COD IN FOOD WASTE PRETREATMENT WITH FUNGAL MASH

Figure 5.3 shows that the sCOD concentration increased significantly in the first 4 h of the pretreatment of FW with the fungal mash and commercial enzymes, respectively. The highest sCOD concentration of 164.7 ± 16.7 g/L was obtained in the FW pretreatment with the fungal mash versus 128.5 ± 2.9 g/L with commercial enzymes. Moreover, 64.3 ± 8.9% and 52 ± 4.9% reduction in volatile suspended solids were achieved at the end of the FW pretreatments with the fungal mash and commercial enzymes, respectively. This indicates that the FW pretreatment using the fungal mash itself would lead to a volume reduction of more than 64% within 24 h.

## 5.3.3 ANAEROBIC DIGESTION OF FOOD WASTE PRETREATED WITH FUNGAL MASH

To maximize volume reduction and energy recovery from FW, the pretreated FW subsequently underwent anaerobic digestion. Figure 5.4 shows the cumulative

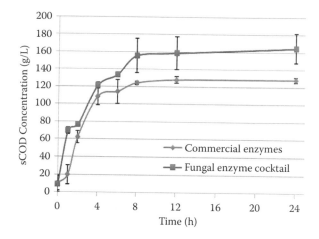

**FIGURE 5.3** Effect of enzymatic pretreatment on soluble COD production from FW. Each data point is the average of triplicate measurements from duplicate experiments and the error bars represent the standard deviations. (Reprinted from *Bioresource Technology*, 183, Uçkun, K.E., A.P. Trzcinski, and Y. Liu., Enhancing the hydrolysis and methane production potential of mixed food wastes by a cost-effective enzymatic pretreatment. 47–52, Copyright 2015, with permission from Elsevier. )

methane yield during the anaerobic digestion of pretreated FW with the fungal mash and commercial enzymes, respectively. The methane yields obtained in these two cases were found to be comparable. However, the methane production from the fungal mash pretreated FW was faster than that of commercial enzymes pretreated FW.

The pretreatments provided almost a full conversion to biogas in the first 13 days. Later, the bioconversion to biogas slowed down. Another advantage of the process is that an anaerobic digester with a short residence time (about 2 weeks) would be required, reducing the capital costs significantly in full scale application. Untreated FW had a lower gas production, indicating that a large fraction of biopolymers in FW without pretreatment would not be readily biodegradable. This shows the importance of proper pretreatment in the anaerobic digestion of FW. Figure 5.4 provides direct evidence that the fungal mash produced *in situ* can significantly enhance the hydrolysis of FW, and improve the efficiency of subsequent anaerobic digestion. In addition to biogas recovery, it should also be pointed out that 80.4 ± 3.5% reduction of volatile solids was achieved after anaerobic digestion of FW pretreated with the fungal mash. The integrated FW pretreatment followed by anaerobic digestion appears to be a promising option for better food waste management in terms of energy recovery and volume reduction.

The experimental data presented in Figure 5.4 were fitted into the Gompertz equation, and the constants estimated are summarized in Table 5.4. A shorter lag phase of 8 h was observed in the anaerobic digestion of FW pretreated with the fungal mash, whereas 12 h for commercial enzymes and 16 h for untreated FW were required. The anaerobic digestion of FW pretreated with the fungal mash was about

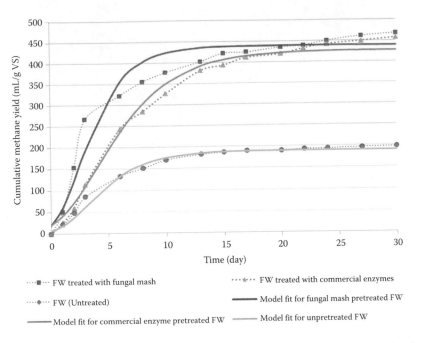

**FIGURE 5.4** Effect of enzymatic pretreatment on cumulative methane production from FW. Results are the average of two replicates. Error was within ± 5 mL/g VS. Solid lines indicate the simulation of the experimental data using Gompertz equation. (Reprinted from *Bioresource Technology*, 183, Uçkun, K.E., A.P. Trzcinski, and Y. Liu., Enhancing the hydrolysis and methane production potential of mixed food wastes by a cost-effective enzymatic pretreatment. 47–52, Copyright 2015, with permission from Elsevier.)

---

## TABLE 5.4
### The Parameters Estimated for the Anaerobic Digestion of FW Pretreated with Different Methods

|                                           | FW without Pretreatment | FW Pretreated with Commercial Enzymes | FW Pretreated with Fungal Mash |
|-------------------------------------------|-------------------------|---------------------------------------|--------------------------------|
| $\lambda$ (hours)                         | 16                      | 12                                    | 8                              |
| $R_m$ (mL $CH_4$/g VS h)                  | 1.1                     | 2.0                                   | 3.8                            |
| $B_0$ (mL $CH_4$/g VS)                    | 190                     | 428                                   | 440                            |
| Experimental yield (mL $CH_4$/g VS)       | 197.9                   | 457.3                                 | 468.2                          |

*Source:* Reprinted from *Bioresource Technology*, 183, Uçkun, K.E., A.P. Trzcinski, and Y. Liu., Enhancing the hydrolysis and methane production potential of mixed food wastes by a cost-effective enzymatic pretreatment. 47–52, Copyright 2015, with permission from Elsevier.

*Note:* $\lambda$: lag phase time, $R_m$: maximum methane production rate, $B_0$: estimated ultimate cumulative methane yield.

1.9 times and 3.5 times faster than the pretreatment with commercial enzymes and untreated FW, respectively.

## 5.4 CONCLUSIONS AND FUTURE PROSPECTS

A fungal mash rich in glucoamylase and protease was produced from cake waste and was applied for enzymatic hydrolysis of mixed FW. The enzymatic pretreatment using this fungal mash was shown to be more efficient than commercial enzymes. The biomethane yield and production rate from FW pretreated with the fungal mash were found to be 3.5 times higher than without pretreatment. The overall volatile suspended solid destruction in the process was $80.4 \pm 3.5\%$. These results showed that direct use of the fungal mash produced *in situ* without any purification steps is a promising option for FW treatment.

In 2015, it was reported that the fungal mash rich in hydrolytic enzymes was *in situ* produced from cake waste (as described in previous chapters), and was further applied for the hydrolysis of MFWs and the subsequent production of ethanol from the hydrolyzate (Uçkun and Liu, 2015). The activities of the hydrolytic enzymes in the fungal mash were determined as follows (in U/g dry fungal mash): glucoamylase $165 \pm 16$, xylanase $112 \pm 7$, β-glucosidase $101 \pm 10$, α-amylase 4, protease $13 \pm 3$, and cellulase $12 \pm 3$. The hydrolysis was carried out in a 3 L bioreactor with 1 L working volume at 60°C, 500 rpm. Blended domestic FW (180 g dry) was inoculated with the fungal mash (7.7 g dry). Then, the total volume of the blend was adjusted to 1 L by adding distilled water. A glucose concentration of 119 g/L was achieved after 12 hours hydrolysis, and it further increased to 127 g/L after 24 hours. A well-balanced nutrient biomedium containing 127 g/L glucose and 1.8 g/L FAN was produced from the enzymatic pretreatment of food wastes. Using this solution as sole fermentation feedstock, 58 g/L of ethanol corresponding to an ethanol yield of 0.5 g/g glucose was obtained within 32 hours. It was demonstrated that the pretreatment of MFW with the fungal mash is an effective option for food waste saccharification and bioethanol production.

In another recent study by Yin et al. (2016), the *in situ*-produced fungal mash rich in hydrolytic enzymes was used for the pretreatment of activated sludge, FW, and their mixture prior to anaerobic digestion. Activated sludge and FW were first concentrated by centrifugation at 7000 rpm and 4°C for 5 minutes. The harvested sludge or FW was adjusted to a final solid concentration of 10 g/L total solid, to which 2 g/L fungal mash was added. Mixed activated sludge and FW were blended in a 1:1 ratio by mass. The enzymatic pretreatment of activated sludge mixed with FW resulted in the production of 3.72 g/L glucose and 51 mg/L FAN, equivalent to 7.65 g/L sCOD within 24 hours. After pretreatment of activated sludge and FW by fungal mash, 19.1% and 21.4% of VS reduction were achieved, respectively. The mixed activated sludge and FW pretreated with fungal mash exhibited an ultimate methane production yield of $600.5 \pm 4.7$ mL $CH_4$/g VS. The biomethane yield of mixed waste pretreated with fungal mash was found to be 2.5 times higher than raw activated sludge without pretreatment, with a further VS reduction of 34.5%. This suggests a total VS reduction of 54.3% in the anaerobic system with the pretreatment by fungal mash. It was also shown that the co-digestion process was kinetically much faster than sole activated sludge or FW digestion. These recent studies show that FW can be converted into biofuels (ethanol, biogas) or into a complex enzyme cocktail that has high value to hydrolyze complex waste such as waste activated sludge. Future work should focus on other waste from the food industry or agricultural residue and fine tune the enzyme cocktail production to suit each individual application. The enzyme cocktail could also be tested on other wastewater containing dyes, phenolic compounds, pesticides, melanoidins, emerging pharmaceuticals, and contaminants.

# References

Acourene, S., K. Djafri, A. Ammouche, L. Amourache, A. Djidda, M. Tama, and B. Taleb. 2011. Utilisation of the date wastes as substrate for the production of baker's yeast and citric acid. *Biotechnology Advances* 10 (6):488–497.

Acourene, S., and A. Ammouche. 2012. Optimization of ethanol, citric acid, and α-amylase production from date wastes by strains of *Saccharomyces cerevisiae, Aspergillus niger,* and *Candida guilliermondii. Journal of Industrial Microbiology and Biotechnology* 39 (5):759–766.

Ado, S.A., G.B. Olukotun, J.B. Ameh, and A. Yabaya. 2009. Bioconversion of cassava starch to ethanol in a simultaneous saccharification and fermentation process by co-cultures of *Aspergillus niger* and *Saccharomyces cerevisiae. Science World Journal* s4:19–22.

Afifi, M.M. 2011. Effective technological pectinase and cellulase by *Saccharomyces cerevisiae* utilizing food wastes for citric acid production. *Life Science Journal* 8 (2):405–413.

Afify, M.M., T.M. Abd El-Ghany, and M.M. Alawlaqi. 2011. Microbial utilization of potato wastes for protease production and their using as biofertilizer. *Australian Journal of Basic and Applied Sciences* 5 (7):308–315.

Akao, S., H. Tsuno, and J. Cheon. 2007. Semi-continuous L-lactate fermentation of garbage without sterile condition and analysis of the microbial structure. *Water Research* 41 (8):1774–1780.

Akiyama, M., Y. Taima, and Y. Doi. 1992. Production of poly(3-hydroxyalkanoates) by a bacterium of the genus *Alcaligenes* utilizing long-chain fatty acids. *Applied Microbiology and Biotechnology* 37 (6):698–701.

Alben, E., and O. Erkmen. 2004. Production of citric acid from a new substrate, undersized semolina, by *Aspergillus niger. Food Technology Biotechnology* 42 (1).19–22.

Alkan, H., Z. Baysal, F. Uyar, and M. Dogru. 2007. Production of lipase by a newly isolated *Bacillus coagulans* under solid-state fermentation using melon wastes. *Applied Biochemistry and Biotechnology* 136 (2):183–192.

Ángel Siles López, J., Q. Li, and I.P. Thompson. 2010. Biorefinery of waste orange peel. *Critical Reviews in Biotechnology* 30 (1):63–69.

Angumeenal, A.R., and D. Venkappayya. 2013. An overview of citric acid production. *LWT — Food Science and Technology* 50 (2):367–370.

Anto, H., U.B. Trivedi, and K.C. Patel. 2006. Glucoamylase production by solid-state fermentation using rice flake manufacturing waste products as substrate. *Bioresource Technology* 97 (10):1161–1166.

AOAC. 2001. *Official Methods of Analysis.* Washington, DC: Association of Official Analytical Chemists.

APHA-WPCF. 1998. *Standard Methods for the Examination of Water and Wastewater.* 20th ed. Washington, DC: American Public Health Association.

Arooj, M.F., S.K. Han, S.H. Kim, D.H. Kim, and H.K. Shin. 2008. Continuous biohydrogen production in a CSTR using starch as a substrate. *International Journal of Hydrogen Energy* 33 (13):3289–3294.

Aye, L., and E.R. Widjaya. 2006. Environmental and economic analysis of waste disposal options for traditional markets in Indonesia. *Waste Management* 26:1180–1191.

Bahcegul, E., Tatli, E., Haykir, N.I., Apaydin, S., and U. Bakir. 2011. Selecting the right blood glucose monitor for the determination of glucose during the enzymatic hydrolysis of corncob pretreated with different methods. *Bioresource Technology* 102:9646–9652.

Bajaj, A., P. Lohan, P.N. Jha, and R. Mehrotra. 2010. Biodiesel production through lipase catalyzed transesterification: An overview. *Journal of Molecular Catalysis B: Enzymatic* 62 (1):9–14.

Ballesteros, M., J.M. Oliva, P. Manzanares, M.J. Negro, and I. Ballesteros. 2002. Ethanol production from paper materials using a simultaneous saccharification and fermentation system in a fed-batch basis. *World Journal of Microbiology and Biotechnology* 18 (6): 559–561.

Banks, C.J., M. Chesshire, S. Heaven, and R. Arnold. 2011. Anaerobic digestion of source-segregated domestic food waste: Performance assessment by mass and energy balance. *Bioresource Technology* 102 (2):612–620.

Bansal, N., R. Tewari, R. Soni, and S.K. Soni. 2012. Production of cellulases from *Aspergillus niger* NS-2 in solid state fermentation on agricultural and kitchen waste residues. *Waste Management* 32:1341–1346.

Barrington, S., and J. Kim. 2008. Response surface optimization of medium components for citric acid production by *Aspergillus niger* NRRL 567 grown in peat moss. *Bioresoure Technology* 99:368–377.

Berovic, M., and H. Ostroversnik. 1997. Production of *Aspergillus niger* pectolytic enzymes by solid state bioprocessing of apple pomace. *Journal of Biotechnology* 53 (1):47–53.

BEST. 2013. Bioethanol from orange in Spain EU 2013 [cited 22 April 2013].

Bialy, H.E., O.M. Gomaa, and K.S. Azab. 2011. Conversion of oil waste to valuable fatty acids using oleaginous yeast. *World Journal of Microbiology and Biotechnology* 27 (12):2791–2798.

Botella, C., I. de Orya, C. Webbb, D. Cantero, and A. Blandino. 2005. Hydrolytic enzyme production by *Aspergillus awamori* on grape pomace. *Biochemical Engineering Journal* 26 (2–3):100–106.

Botella, C., A. Diaz, I. de Ory, C. Webb, and A. Blandino. 2007. Xylanase and pectinase production by *Aspergillus awamori* on grape pomace in solid state fermentation. *Process Biochemistry* 42:98–101.

Bozell, J.J., and G.R. Petersen. 2010. Technology development for the production of biobased products from biorefinery carbohydrates—The US Department of Energy's "top 10" revisited. *Green Chemistry* 12 (4):539–554.

Castilho, L.R., D.A. Mitchell, and M.G. Denise. 2009. Production of polyhydroxyalkanoates (PHAs) from waste materials and by-products by submerged and solid-state fermentation. *Bioresource Technology* 100 (23):5996–6009.

Cavaleiro, A.J., T. Ferreira, F. Pereira, G. Tommaso, and M.M. Alves. 2013. Biochemical methane potential of raw and pre-treated meat-processing wastes. *Bioresoure Technology* 129:519–525.

Cekmecelioglu, D., A. Demirci, R.E. Graves, and N.H. Davitt. 2005. Applicability of optimized in-vessel food waste composting for windrow systems. *Biosystems Engineering* 91:479–486.

Cekmecelioglu, D., and O.N. Uncu. 2013. Kinetic modeling of enzymatic hydrolysis of pre-treated kitchen wastes for enhancing bioethanol production. *Waste Management* 33 (3):735–739.

Chandel, A.K., G. Chandrasekhar, M.B. Silva, and S. Silvério Da Silva. 2012. The realm of cellulases in biorefinery development. *Critical Reviews in Biotechnology* 32 (3):187–202.

Chen, G., and M.K. Patel. 2012. Plastics derived from biological sources: Present and future: A technical and environmental review. *Chemical Reviews* 112 (4):2082–2099.

Chen, Y., X. Li, X. Zheng, and D. Wang. 2013. Enhancement of propionic acid fraction in volatile fatty acids produced from sludge fermentation by the use of food waste and *Propionibacterium acidipropionici*. *Water Research* 47 (2):615–622.

Chen, Y., B. Xiao, J. Chang, Y. Fu, P. Lv, and X. Wang. 2009. Synthesis of biodiesel from waste cooking oil using immobilized lipase in fixed bed reactor. *Energy Conversion and Management* 50:668–673.

Chien, C., and L. Ho. 2008. Polyhydroxyalkanoates production from carbohydrates by a genetic recombinant *Aeromonas* sp. *Letters in Applied Microbiology* 47 (6):587–593.

Cho, J.K., S.C. Park, and H.N. Chang. 1995. Biochemical methane potential and solid state anaerobic digestion of Korean food wastes. *Bioresource Technology* 52 (3):245–253.

Chu, C., Y. Li, K. Xu, Y. Ebie, Y. Inamori, and H. Kong. 2008. A pH- and temperature-phased two-stage process for hydrogen and methane production from food waste. *International Journal of Hydrogen Energy* 33 (18):4739–4746.

Chutmanop, J., S. Chuichulcherm, Y. Chisti, and P. Srinophakun. 2008. Protease production by *Aspergillus oryzae* in solid-state fermentation using agroindustrial substrates. *Journal of Chemical Technology and Biotechnology* 83 (7):1012–1018.

Cira, L.A., S. Huerta, G.M. Hall, and K. Shirai. 2002. Pilot scale lactic acid fermentation of shrimp wastes for chitin recovery. *Process Biochemistry* 37 (12):1359–1366.

Citrotechno. 2013. Second generation bioethanol, 22.04.2013 2013 [cited 01 May 2013].

Colen, G., R.G. Junqueira, and T. Moraes-Santos. 2006. Isolation and screening of alkaline lipase-producing fungi from Brazilian savanna soil. *Journal of Microbiology and Biotechnology* 22:881–885.

Contesini, F.J., D.B. Lopes, G.A. MacEdo, M.D.G. Nascimento, and P.D.O. Carvalho. 2010. Aspergillus sp. lipase: Potential biocatalyst for industrial use. *Journal of Molecular Catalysis B: Enzymatic* 67 (3–4):163–171.

Couto, S.R., and M.A. Sanromán. 2006. Application of solid-state fermentation to food industry—A review. *Journal of Food Engineering* 76 (3):291–302.

Dai, X., N. Duan, B. Dong, and L. Dai. 2013. High-solids anaerobic co-digestion of sewage sludge and food waste in comparison with mono digestions: Stability and performance. *Waste Management* 33 (2):308–316.

De Castro, A.M., T.V. De Andréa, D.F. Carvalho, M.M.P. Teixeira, L. Dos Reis Castilho, and D.M.G. Freire. 2011. Valorization of residual agroindustrial cakes by fungal production of multienzyme complexes and their use in cold hydrolysis of raw starch. *Waste and Biomass Valorization* 2 (3):291–302.

Dhillon, G.S., S.K. Brar, M. Verma, and R.D. Tyagi. 2011. Enhanced solid-state citric acid bio-production using apple pomace waste through surface response methodology. *Journal of Applied Microbiology* 110 (4):1045–1055.

Dhillon, G.S., S. Kaura, S.K. Brara, and M. Vermac. 2012. Potential of apple pomace as a solid substrate for fungal cellulase and hemicellulase bioproduction through solid-state fermentation. *Industrial Crops and Products* 38 (1):6–13.

Díaz, A.B., I. de Ory, I. Caro, and A. Blandino. 2012. Enhance hydrolytic enzymes production by *Aspergillus awamori* on supplemented grape pomace. *Food and Bioproducts Processing* 90 (1):72–78.

Dijkstra, A.J. 2010. Enzymatic degumming. *European Journal of Lipid Science and Technology* 112 (11):1178–1189.

Dominguez, A., F.J. Deive, M.A. Sanroman, and M.A. Longo. 2010. Biodegradation and utilization of waste cooking oil by *Yarrowia lipolytica* CECT 1240. *European Journal of Lipid Science and Technology* 112 (11):1200–1208.

Dorado, M.P., S.K.C. Lin, A. Koutinas, C. Du, R. Wang, and C. Webb. 2009. Cereal-based biorefinery development: Utilisation of wheat milling by-products for the production of succinic acid. *Journal of Biotechnology* 143 (1):51–59.

Dos Santos, T.C., D.P.P. Gomes, R.C.F. Bonomo, and M. Franco. 2012. Optimisation of solid state fermentation of potato peel for the production of cellulolytic enzymes. *Food Chemistry* 133:1299–1304.

Du, C., S.K.C. Lin, A. Koutinas, R. Wang, P. Dorado, and C. Webb. 2008. A wheat biorefin-ing strategy based on solid-state fermentation for fermentative production of succinic acid. *Bioresource Technology* 99:8310–8315.

Du, G., and J. Yu. 2002. Green technology for conversion of food scraps to biodegradable thermoplastic polyhydroxyalkanoates. *Environmental Science and Technology* 36 (24):5511–5516.

Du, G., L.X.L. Chen, and J. Yu. 2004. High-efficiency production of bioplastics from biode-gradable organic solids. *Journal of Polymers and the Environment* 12 (2):89–94.

Ducroo, P. 1987. Improvements relating to the production of glucose syrups and purified starches from wheat and other cereal starches containing pentosans. EU Patent EP 228,732; *Chemistry Abstracts 108*, page 4704d.

E-fuel. 2013. *Earth's First Home Ethanol System* (2009), 22.04.2013 2009 [cited 01 May 2013].

Edwinoliver, N.G., K. Thirunavukarasu, R.B. Naidu, M.K. Gowthaman, T. Nakajima Kambe, and N.R. Kamini. 2010. Scale up of a novel tri-substrate fermentation for enhanced production of *Aspergillus niger* lipase for tallow hydrolysis. *Bioresource Technology* 101:6791–6796.

Effendi, A., H. Gerhauser, and A.V. Bridgwater. 2008. Production of renewable phenolic res-ins by thermochemical conversion of biomass: A review. *Renewable and Sustainable Energy Reviews* 12 (8):2092–2116.

El-Fadel, M., A.N. Findikakis, and J.O. Leckie. 1997. Environmental impacts of solid waste landfilling. *Journal of Environmental Management* 50 (1):1–25.

El-Mashad, H.M., J.A. McGarvey, and R. Zhang. 2008. Performance and microbial analy-sis of anaerobic digesters treating food waste and dairy manure. *Biology Engineering* 1:233–242.

Elayaraja, S., T. Velvizhi, V. Maharani, P. Mayavu, S. Vijayalakshmi, and T. Balasubramanian. 2011. Thermostable alpha-amylase production by *Bacillus firmus* CAS 7 using potato peel as a substrate. *African Journal of Biotechnology* 10 (54):11235–11238.

Elbeshbishy, E., H. Hafez, B.R. Dhar, and G. Nakhla. 2011. Single and combined effect of var-ious pretreatment methods for biohydrogen production from food waste. *International Journal of Hydrogen Energy* 36 (17):11379–11387.

Ellaiah, P., K. Adinarayana, Y. Bhavani, P. Padmaja, and B. Srinivasulu. 2002. Optimization of process parameters for glucoamylase production under solid state fermentation by a newly isolated *Aspergillus* species. *Process Biochemistry* 38 (4):615–620.

Energy-Enviro-Finland. 2013. First plants producing ethanol from food waste in progress (27.02.2007). EnergyEnviroFinland, 22.04.2013 2013 [cited 23 April 2013].

Esakkiraj, P., R. Usha, A. Palavesam, and G. Immanuel. 2012. Solid-state production of esterase using fish processing wastes by *Bacillus altitudinis* AP-MSU. *Food and Bioproducts Processing* 90:370–376.

Eshtaya, M.K., N.A. Rahman, and M.A. Hassan. 2013. Bioconversion of restaurant waste into Polyhydroxybutyrate (PHB) by recombinant E. coli through anaerobic digestion. *International Journal of Environment and Waste Management* 11 (1):27–37.

Esteban, M.B., A.J. Garcia, P. Ramos, and M.C. Marquez. 2007. Evaluation of fruit, vegetable and fish wastes as alternative feedstuffs in pig diets. *Waste Management* 27:193–200.

Expert Committee on Food Additives. 1967. Lactic acid. *WHO/Food Additives* 29:144–148.

Fakas, S., S. Papanikolaou, M. Galiotou-Panayotou, M. Komaitis, and G. Aggelis. 2008a. Organic nitrogen of tomato waste hydrolysate enhances glucose uptake and lipid accumulation in *Cunninghamella echinulata. Journal of Applied Microbiology* 105:1062–1070.

Fakas, S., M. Certik, S. Papanikolaou, G. Aggelis, M. Komaitis, and M. Galiotou-Panayotou. 2008b. Linolenic acid production by *Cunninghamella echinulata* growing on complex organic nitrogen sources. *Bioresource Technology* 99:5986–5990.

Fakas, S., A. Makri, M. Mavromati, M. Tselepi, and G. Aggelis. 2009. Fatty acid composition in lipid fractions lengthwise the mycelium of *Mortierella isabellina* and lipid production by solid state fermentation. *Bioresource Technology* 100:6118–6120.

Falony, G., J.C. Armas, J.C.D. Mendoza, and J.L.M. Hernández. 2006. Production of extracellular lipase from *Aspergillus niger* by solid-state fermentation. *Food Technology Biotechnology* 44 (2):235–240.

Fan, Y., C. Zhou, and X. Zhu. 2009. Selective catalysis of lactic acid to produce commodity chemicals. *Catalysis Reviews—Science and Engineering* 51 (3):293–324.

FAO. 2012. *Towards the Future We Want: End Hunger and Make the Transition to Sustainable Agricultural and Food Systems*. Rome: Food and Agriculture Organization of the United Nations.

Forster-Carneiro, T., M. Pérez, and L.I. Romero. 2008. Influence of total solid and inoculum contents on performance of anaerobic reactors treating food waste. *Bioresource Technology* 99 (15):6994–7002.

Gajalakshmi, S., and S.A. Abbasi. 2008. Solid waste management by composting: State of the art. *Critical Reviews in Environmental Science and Technology* 38 (5):311–400.

Gao, C., C. Ma, and P. Xu. 2011. Biotechnological routes based on lactic acid production from biomass. *Biotechnology Advances* 29 (6):930–939.

Garzon, C.G., and R.A. Hours. 1992. Citrus waste: An alternative substrate for pectinase production in solid-state culture. *Bioresource Technology* 39 (1):93–95.

Giese, E.C., R.F.H. Dekker, and A.M. Barbosa. 2008. Orange bagasse as substrate for the production of pectinase and laccase by *Botryosphaeria rhodina* MAMB-05 in submerged and solid state fermentation. *BioResources* 3 (2):335–345.

Gomez, X., M. Cuetos, J. Cara, A. Moran, and A. Garcia. 2005. Anaerobic co-digestion of primary sludge and the fruit and vegetable fraction of the municipal solid wastes: Conditions for mixing and evaluation of the organic loading rate. *Renewable Energy* 31 (12):2017–2024.

Gómez, X., C. Fernández, J. Fierro, M.E. Sánchez, A. Escapa, and A. Morán. 2011. Hydrogen production: Two stage processes for waste degradation. *Bioresource Technology* 102:8621–8627.

Gullon, B., G. Garrote, J.L. Alonso, and J.C. Parajo. 2007. Production of L-lactic acid and oligomeric compounds from apple pomace by simultaneous saccharification and fermentation: A response surface methodology assessment. *Journal of Agricultural and Food Chemistry* 55 (14):5580–5587.

Gullon, B., R. Yanez, J.L. Alonso, and J.C. Parajo. 2008. L-Lactic acid production from apple pomace by sequential hydrolysis and fermentation. *Bioresource Technology* 99:308–319.

Gunaseelan, V.N. 2004. Biochemical methane potential of fruits and vegetable solid waste feedstocks. *Biomass and Bioenergy* 26 (4):389–399.

Gupta, N., V. Shai, and R. Gupta. 2007. Alkaline lipase from a novel strain *Burkholderia multivorans*: Statistical medium optimization and production in a bioreactor. *Process Biochemistry* 42 (2):518–526.

Gupta, R.K., D. Prasad, J. Sathesh, R.B. Naidu, N.R. Kamini, S. Palanivel, and M.K. Gowthaman. 2012. Scale-up of an alkaline protease from *Bacillus pumilus* MTCC 7514 utilizing fish meal as a sole source of nutrients. *Journal of Microbiology and Biotechnology* 22 (9):1230–1236.

Gurpreet, S.D., K.B. Satinder, V. Mausam, and T. Rajeshwar. 2011. Recent advances in citric acid bio-production and recovery. *Food Bioprocess Technology* 4 (4):505–529.

Gustafsson, U., W. Wills, and A. Draper. 2011. Food and public health: Contemporary issues and future directions. *Critical Public Health* 21 (4):385–393.

Hafid, H.S., N.A.A. Rahman, S. Abd-Aziz, and M.A. Hassan. 2011. Enhancement of organic acids production from model kitchen waste via anaerobic digestion. *African Journal of Biotechnology* 10 (65):14507–14515.

Hafid, H.S., N.A.A. Rahman, F.N. Omar, P.L. Yee, S. Abd-Aziz, and M.A. Hassan. 2010. A comparative study of organic acids production from kitchen wastes and simulated kitchen waste. *Australian Journal of Basic and Applied Sciences* 4 (4):639–645.

Hafuka, A., K. Sakaida, H. Satoh, M. Takahashi, Y. Watanabe, and S. Okabe. 2011. Effect of feeding regimens on polyhydroxybutyrate production from food wastes by *Cupriavidus necator*. *Bioresource Technology* 102 (3):3551–3553.

Hallenbeck, P.C., and D. Ghosh. 2009. Advances in fermentative biohydrogen production: The way forward? *Trends in Biotechnology* 27 (5):287–297.

Hamdy, H.S. 2013. Citric acid production by *Aspergillus niger* grown on orange peel medium fortified with cane molasses. *Annals of Microbiology* 63 (1):267–278.

Han, S.K., and H.S. Shin. 2004. Biohydrogen production by anaerobic fermentation of food waste. *International Journal of Hydrogen Energy* 29 (6):569–577.

Hara, A., and N.S. Radin. 1978. Lipid extraction of tissues with a low toxicity solvent. *Analytical Biochemistry* 90:420–426.

Hari Krishna, H., T. Reddy Janardhan, and G.V. Chowdary. 2001. Simultaneous saccharification and fermentation of lignocellulosic wastes to ethanol using a thermotolerant yeast. *Bioresource Technology* 77:193–196.

He, M., Y. Sun, D. Zou, H. Yuan, B. Zhu, X. Li, and Y. Pang. 2012a. Influence of temperature on hydrolysis acidification of food waste. *Procedia Environmental Sciences* 16:85–94.

He, M.X., H. Feng, F. Bai, Y. Li, X. Liu, and Y.Z. Zhang. 2009. Direct production of ethanol from raw sweet potato starch using genetically engineered *Zymomonas mobilis*. *African Journal of Biotechnology* 3:721–726.

He, Y., D.M. Bagley, K.T. Leung, S.N. Liss, and B. Liao. 2012b. Recent advances in membrane technologies for biorefining and bioenergy production. *Biotechnology Advances* 30 (4):817–858.

Heo, N.H., S.C. Park, and H. Kang. 2004. Effects of mixture ratio and hydraulic retention time on single-stage anaerobic co-digestion of food waste and waste activated sludge. *Journal of Environmental Science and Health—Part A Toxic/Hazardous Substances and Environmental Engineering* 39 (7):1739–1756.

Hirai, Y., M. Murata, S. Sakai, and H. Takatsuki. 2001. Life cycle assessment on food waste management and recycling. *Waste Management Resources* 12 (5):219–228.

Hong, K., C.L. Yun, Y.K. Sui, H.L. Kin, H.L. Wai, H. Chua, and P.H.F. Yu. 2000. Construction of recombinant *Escherichia coli* strains for polyhydroxybutyrate production using soy waste as nutrient. *Applied Biochemistry and Biotechnology—Part A Enzyme Engineering and Biotechnology* 84–86:381–390.

Hong, Y.S., and H.H. Yoon. 2011. Ethanol production from food residues. *Biomass Bioenergy* 35 (7):3271–3275.

Hours, R.A., C.E. Voget, and R.J. Ertola. 1988. Some factors affecting pectinase production from apple pomace in solid-state cultures. *Biological Wastes* 24 (2):147–157.

Huang, C., X.F. Chen, L. Xiong, X. Chen, L.L. Ma, and Y. Chen. 2013. Single cell oil production from low-cost substrates: The possibility and potential of its industrialization. *Biotechnology Advances* 31:129–139.

Idris, A., and W. Suzana. 2006. Effect of sodium alginate concentration, bead diameter, initial pH and temperature on lactic acid production from pineapple waste using immobilized *Lactobacillus delbrueckii*. *Process Biochemistry* 41 (5):1117–1123.

Imandi, S.B., V.V.R. Bandaru, S.R. Somalanka, S.R. Bandaru, and H.R. Garapati. 2008. Application of statistical experimental designs for the optimization of medium constituents for the production of citric acid from pineapple waste. *Bioresource Technology* 99 (10):4445–4450.

International Renewable Energy Agency. 2013. Production of bio-ethylene: IRENA-ETSAP.

Jamrath, T., C. Lindner, M.K. Popovic, and R. Bajpai. 2012. Amylase and protease production by *B. caldolyticus*. *Food Technology and Biotechnology* 50 (3):355–361.

Jang, Y., B. Kim, J.H. Shin, Y.J. Choi, S. Choi, C.W. Song, J. Lee, H.G. Park, and S.Y. Lee. 2012. Bio-based production of C2-C6 platform chemicals. *Biotechnology and Bioengineering* 109 (10):2437–2459.

Japan for Sustainability. 2013. Public-Private-Academic Partnership in Kyoto to Convert Municipal Solid Waste into Ethanol. In Japan for Sustainability. http://www.japanfs.org/en/pages/031484.html.

Jawad, A.H., A.F.M. Alkarkhi, O.C. Jason, A.M. Easa, and N.A. Nik Norulaini. 2013. Production of the lactic acid from mango peel waste—Factorial experiment. *Journal of King Saud University—Science* 25 (1):39–45.

Jellouli, K., A. Bayoudh, L. Manni, R. Agrebi, and M. Nasri. 2008. Purification, biochemical and molecular characterization of a metalloprotease from *Pseudomonas aeruginosa* MN7 grown on shrimp wastes. *Applied Microbiology and Biotechnology* 79 (6):989–999.

Jensen, J.W., C. Felby, and H. Jørgensen. 2011. Cellulase hydrolysis of unsorted MSW. *Applied Biochemistry and Biotechnology* 165:1799–1811.

Jeong, S., Y. Kim, and D. Lee. 2012. Ethanol production by co-fermentation of hexose and pentose from food wastes using *Saccharomyces coreanus* and *Pichia stipitis*. *Korean Journal of Chemical Engineering* 29 (8):1038–1043.

Jørgensen, H., J.B. Kristensen, and C. Felby. 2007. Enzymatic conversion of lignocellulose into fermentable sugars: Challenges and opportunities. *Biofuels Bioproducts Biorefinery* 1 (2):119–134.

Jung, Y.M., and Y.H. Lee. 2000. Utilization of oxidative pressure for enhanced production of polyhydroxybutyrate and poly-3-hydroxybutyrate-3- hydroxyvalerate in *Ralstonia eutropha*. *Bioscience and Bioengineering* J90 (3):266–270.

Kamzolova, S.V., I.G. Morgunov, and T.V. Finogenova. 2008. Microbiological production of citric and isocitric acid from sunflower oil. *Food Technology and Biotechnology* 46 (1):51–59.

Karasu Yalçin, S. 2012. Enhancing citric acid production of *Yarrowia lipolytica* by mutagenesis and using natural media containing carrot juice and celery byproducts. *Food Science and Biotechnology* 21 (3):867–874.

Karthikeyan, A., and N. Sivakumar. 2010. Citric acid production by Koji fermentation using banana peel as a novel substrate. *Bioresource Technology* 101: 5552–5556.

Kashyap, D.R., P.K. Vohra, S. Chopra, and R. Tewari. 2001. Applications of pectinases in the commercial sector: A review. *Bioresource Technology* 77 (3):215–227.

Kastner, V., W. Somitsch, and W. Schnitzhofer. 2012. The anaerobic fermentation of food waste: A comparison of two bioreactor systems. *Journal of Cleaner Production* 34:82–90.

Katami, T., A. Yasuhara, and T. Shibamoto. 2004. Formation of dioxins from incineration of foods found in domestic garbage. *Environmental Science and Technology* 38 (4):1062–1065.

Kawa-Rygielska, J., W. Pietrzak, and A. Czubaszek. 2012. Characterization of fermentation of waste wheat-rye bread mashes with the addition of complex enzymatic preparations. *Biomass Bioenergy* 44:17–22.

Kayhanian, M., and D. Rich. 1995. Pilot-scale high solids thermophilic anaerobic digestion of municipal solid waste with an emphasis on nutrient requirements. *Biomass Bioenergy* 8 (6):433–444.

Khandeparkar, R.D.S., and N.B. Bhosle. 2006. Isolation, purification and characterization of the xylanase produced by *Arthrobacter* sp. MTCC 5214 when grown in solid-state fermentation. *Enzyme and Microbial Technology* 39 (4):732–742.

Khosravi-Darani, K., H.R. Falahatpishe, and M. Jalali. 2008. Alkaline protease production on date waste by an alkalophilic *Bacillus* sp. 2–5 isolated from soil. *African Journal of Biotechnology* 7 (10):1536–1542.

Khuri, A.I., and J.A. Cornell. 1987. *Response Surfaces: Design and Analysis*. New York: Marcel Dekker.

Kim, D., S. Kim, H. Kim, M. Kim, and H. Shin. 2011a. Sewage sludge addition to food waste synergistically enhances hydrogen fermentation performance. *Bioresource Technology* 102 (18):8501–8506.

Kim, D.H., and M.S. Kim. 2013. Development of a novel three-stage fermentation system converting food waste to hydrogen and methane. *Bioresource Technology* 127:267–274.

Kim, D.H., S.H. Kim, H.Y. Kim, and H.S. Shin. 2010. Experience of a pilot-scale hydrogen-producing anaerobic sequencing batch reactor (ASBR) treating food waste. *International Journal of Hydrogen Energy* 35:1590–1594.

Kim, D.H., S.H. Kim, and H.S. Shin. 2008a. Hydrogen fermentation of food waste without inoculum addition. *Enzyme and Microbial Technology* 45:181–187.

Kim, H.J., S.H. Kim, Y.G. Choi, G.D. Kim, and H. Chung. 2006a. Effect of enzymatic pretreatment on acid fermentation of food waste. *Journal of Chemical Technology and Biotechnology* 81 (6):974–980.

Kim, J.H., J.C. Lee, and D. Pak. 2011b. Feasibility of producing ethanol from food waste. *Waste Management* 31:2121–2125.

Kim, J.H., J.C. Lee, and D. Pak. 2011c. Feasibility of producing ethanol from food waste. *Waste Management* 31 (9–10):2121–2125.

Kim, J.K., G.H. Han, B.R. Oh, Y.N. Chun, C.Y. Eom, and S.W. Kim. 2008b. Volumetric scale-up of a three stage fermentation system for food waste treatment. *Bioresource Technology* 99:4394–4399.

Kim, J.K., B.R. Oh, Y.N. Chun, and S.W. Kim. 2006b. Effects of temperature and hydraulic retention time on anaerobic digestion of food waste. *Journal of Bioscience and Bioengineering* 102 (4):328–332.

Kim, J.K., B.R. Oh, H. Shin, C. Eom, and S.W. Kim. 2008c. Statistical optimization of enzymatic saccharification and ethanol fermentation using food waste. *Process Biochemistry* 43 (11):1308–1312.

Kim, K.C., S.W. Kim, M.J. Kim, and S.J. Kim. 2005. Saccharification of food wastes using cellulolytic and amylolytic enzymes from *Trichoderma harzianum* FJ1 and its kinetics. *Biotechnology and Bioprocess Engineering* 10:52–59.

Kim, K.I., W.K. Kim, D.K. Seo, I.S. Yoo, E.K. Kim, and H.H. Yoon. 2003. Production of lactic acid from food wastes. *Applied Biochemistry and Biotechnology—Part A Enzyme Engineering and Biotechnology* 107 (105–108):637–647.

Kim, S., and B.E. Dale. 2004. Global potential bioethanol production from wasted crops and crop residues. *Biomass and Bioenergy* 26 (4):361–375.

Kim, S.H., S.K. Han, and H.S. Shin. 2004. Feasibility of biohydrogen production by anaerobic co-digestion of food waste and sewage sludge. *International Journal of Hydrogen Energy* 29:1607–1616.

Koike, Y., M.Z. An, Y.Q. Tang, T. Syo, N. Osaka, S. Morimura, and K. Kida. 2009. Production of fuel ethanol and methane from garbage by high-efficiency two-stage fermentation process. *Journal of Bioscience and Bioengineering* 108 (6):508–512.

Koutinas, A.A., N. Arifeen, R. Wang, and C. Webb. 2007a. Cereal-based biorefinery development: Integrated enzyme production for cereal flour hydrolysis. *Biotechnology and Bioengineering* 97:61–72.

Koutinas, A.A., F. Malbranque, R. Wang, G.M. Campbell, and C. Webb. 2007b. Development of an oat-based biorefinery for the production of L(+)-lactic acid by rhizopus oryzae and various value-added coproducts. *Journal of Agricultural and Food Chemistry* 55 (5):1755–1761.

Koutinas, A.A., A. Vlysidis, D. Pleissner, N. Kopsahelis, I. Lopez Garcia, I.K. Kookos, S. Papanikolaou, T.H. Kwan, and C.S.K. Lin. 2014. Valorization of industrial waste and by-product streams via fermentation for the production of chemicals and biopolymers. *Chemical Society Reviews* 43 (8):2587–2627.

Koutinas, A.A., R. Wang, and C. Webb. 2007c. The biochemurgist—Bioconversion of agricultural raw materials for chemical production. *Biofuels, Bioproducts and Biorefining* 1 (1):24–38.

Krishna, C. 1999. Production of bacterial cellulases by solid state bioprocessing of banana wastes. *Bioresource Technology* 69 (3):231–239.

Kuhad, R.C., R. Gupta, and A. Singh. 2011. Microbial cellulases and their industrial applications. *Enzyme Research* 2011 (1):10.

Kumar, D., V.K. Jain, G. Shanker, and A. Srivastava. 2003. Utilisation of fruits waste for citric acid production by solid state fermentation. *Process Biochemistry* 38 (12):1725–1729.

Kumar, D., R. Verma, and Bhalla T.C. 2010. Citric acid production by *Aspergillus niger* van. Tieghem MTCC 281 using waste apple pomace as a substrate. *Journal of Food Science and Technology* 47 (4):458–460.

Kumar, J.V., A. Shahbazi, and R. Mathew. 1998. Bioconversion of solid food wastes to ethanol. *Analyst* 123 (3):497–502.

Kwon, S.H., and D.H. Lee. 2004. Evaluation of Korean food waste composting with fed-batch operations I: Using water extractable total organic carbon contents (TOCw). *Process Biochemistry* 39 (10):1183–1194.

Kyazze, G., R. Dinsdale, A.J. Guwy, F.R. Hawkes, G.C. Premier, and D.L. Hawkes. 2007. Performance characteristics of two-stage dark fermentative system producing hydrogen and methane continuously. *Biotechnology and Bioengineering* 97 (4):759–770.

Kyle, D.J. 2001. *The Large-Scale Production and Use of a Single-Cell Oil Highly Enriched in Docosahexaenoic Acid.* ACS Symposium Series, Vol. 788, pp. 92–107.

Lam, K.F., C.C.J. Leung, H.M. Lei, and C.S.K. Lin. 2014. Economic feasibility of a pilot-scale fermentative succinic acid production from bakery wastes. *Food and Bioproducts Processing* 92:282–290.

Latif, M.A., A. Ahmad, R. Ghufran, and Z.A. Wahid. 2012. Effect of temperature and organic loading rate on upflow anaerobic sludge blanket reactor and $CH_4$ production by treating liquidized food waste. *Environmental Progress and Sustainable Energy* 31 (1):114–121.

Lau, K.Y., D. Pleissner, and C.S.K. Lin. 2014. Recycling of food waste as nutrients in *Chlorella vulgaris* cultivation. *Bioresource Technology* 170:144–151.

Lee, D., Y. Ebie, K. Xu, Y. Li, and Y. Inamori. 2010a. Continuous $H_2$ and $CH_4$ production from high-solid food waste in the two-stage thermophilic fermentation process with the recirculation of digester sludge. *Bioresource Technology* 101:s42–s47.

Lee, J.P., J.S. Lee, and S.C. Park. 1999. Two-phase methanization of food wastes in pilot scale. *Applied Biochemistry and Biotechnology—Part A Enzyme Engineering and Biotechnology* 77–79:585–593.

Lee, Y.W., and J. Chung. 2010. Bioproduction of hydrogen from food waste by pilot-scale combined hydrogen/methane fermentation. *International Journal of Hydrogen Energy* 35:11746–11755.

Lee, Z., S. Li, P. Kuo, I. Chen, Y. Tien, Y. Huang, C. Chuang, S. Wong, and S. Cheng. 2010b. Thermophilic bio-energy process study on hydrogen fermentation with vegetable kitchen waste. *International Journal of Hydrogen Energy* 35 (24):13458–13466.

Lee, Z.K., S.L. Li, J.S. Lin, Y.H. Wang, P.C. Kuo, and S.S. Cheng. 2008. Effect of pH in fermentation of vegetable kitchen wastes on hydrogen production under a thermophilic condition. *International Journal of Hydrogen Energy* 33:5234–5241.

Leung, C.C.J., A.S.Y. Cheung, A.Y.Z. Zhang, K.F. Lam, and C.S.K. Lin. 2012. Utilisation of waste bread for fermentative succinic acid production. *Biochemical Engineering Journal* 65:10–15.

Li, C., P. Champagne, and B.C. Anderson. 2013. Effects of ultrasonic and thermo-chemical pre-treatments on methane production from fat, oil and grease (FOG) and synthetic kitchen waste (KW) in anaerobic co-digestion. *Bioresource Technology* 130:187–197.

Li, C.L., and H.H.P. Fang. 2007. Fermentative hydrogen production from wastewater and solid wastes by mixed cultures. *Critical Reviews in Environmental Science and Technology* 37 (3):1–39.

Li, H., L. Yang, Y.J. Kim, and S.J. Kim. 2011. Continuous ethanol production by the synchronous saccharification and fermentation using food wastes. *Korean Journal of Chemical Engineering* 28 (4):1085–1089.

Li, N.W., M.H. Zong, and H. Wu. 2009. Highly efficient transformation of waste oil to biodiesel by immobilized lipase from *Penicillium expansum*. *Process Biochemistry* 44 (6):685–688.

Li, Q., J.A. Siles, and I.P. Thompson. 2010. Succinic acid production from orange peel and wheat straw by batch fermentations of *Fibrobacter succinogenes* S85. *Applied Microbiology and Biotechnology* 88 (3):671–678.

Lie, S. 1973. The EBC-ninhydrin method for determination of free alpha amino nitrogen. *Journal of the Institute of Brewing* 79 (1):37–41.

Lin, C.S.K., L.A. Pfaltzgraff, L. Herrero-Davila, E.B. Mubofu, S. Abderrahim, J.H. Clark, A.A. Koutinas, et al. 2013. Food waste as a valuable resource for the production of chemicals, materials and fuels. Current situation and global perspective. *Energy and Environmental Science* 6 (2):426–464.

Lohrasbi, M., M. Pourbafrani, C. Niklasson, and M.J. Taherzadeh. 2010. Process design and economic analysis of a citrus waste biorefinery with biofuels and limonene as products. *Bioresource Technology* 101 (19):7382–7388.

López, J.A., C. da Costa Lázaroa, L. dos Reis Castilho, D.M. Guimarães Freire, and A.M. de Castroc. 2013. Characterization of multienzyme solutions produced by solid-state fermentation of babassu cake, for use in cold hydrolysis of raw biomass. *Biochemical Engineering Journal* 77:231–239.

Lundgren, A., and T. Hjertberg. 2010. Ethylene from renewable resources. In *Surfactants from Renewable Resources* (M. Kjellin and I. Johansson, eds), John Wiley & Sons, Ltd, Chichester, UK. doi: 10.1002/9780470686607.ch6

Luo, G., L. Xie, Z. Zou, W. Wang, and Q. Zhou. 2010. Evaluation of pretreatment methods on mixed inoculum for both batch and continuous thermophilic biohydrogen production from cassava stillage. *Bioresource Technology* 101 (3):959–964.

Ma, H., Q. Wang, D. Qian, L. Gong, and W. Zhang. 2009a. The utilization of acid-tolerant bacteria on ethanol production from kitchen garbage. *Renewable Energy* 34 (6):1466–1470.

Ma, H., Q. Wang, W. Zhang, W. Xu, and D. Zou. 2008. Optimization of the medium and process parameters for ethanol production from kitchen garbage by *Zymomonas mobilis*. *International Journal of Green Energy* 5 (6):480–490.

Ma, K.D., M. Wakisaka, T. Kiuchi, S. Praneetrattananon, S. Morimura, K. Kida, and Y. Shirai. 2007. Repeated-batch ethanol fermentation of kitchen refuse by acid tolerant flocculating yeast under the non-sterilized condition. *Japan Journal of Food Engineering* 8:275–279.

Ma, K., M. Wakisaka, K. Sakai, and Y. Shirai. 2009b. Flocculation characteristics of an isolated mutant flocculent *Saccharomyces cerevisiae* strain and its application for fuel ethanol production from kitchen refuse. *Bioresource Technology* 100 (7):2289–2292.

MacIel, M., C. Ottoni, C. Santos, N. Lima, K. Moreira, and C. Souza-Motta. 2013. Production of polygalacturonases by *Aspergillus* section Nigri strains in a fixed bed reactor. *Molecules* 18 (2):1660–1671.

Mahmood, T., and S.T. Hussain. 2010. Nanobiotechnology for the production of biofuels from spent tea. *African Journal of Biotechnology* 9 (6):858–868.

Marin, J., K.J. Kennedy, and C. Eskicioglu. 2010. Effect of microwave irradiation on anaerobic degradability of model kitchen waste. *Waste Management* 30:1772–1779.

Martínez Sabajanes, M., R. Yáñez, J.L. Alonso, and J.C. Parajó. 2012. Pectic oligosaccharides production from orange peel waste by enzymatic hydrolysis. *International Journal of Food Science and Technology* 47 (4):747–754.

Massanet-Nicolau, J., R. Dinsdale, A. Guwy, and G. Shipley. 2013. Use of real time gas production data for more accurate comparison of continuous single-stage and two-stage fermentation. *Bioresource Technology* 129:561–567.

Md Din, M.F., S.C. Huey, S. Muhd Yunus, M.A. Ahmad, M. Ponraj, Z. Ujang, and C. Shreeshivadasan. 2012. Raw material resource for biodegradable plastic production from cafeteria wastes. *Journal of Scientific and Industrial Research* 71 (8):573–578.

Melikoglu, M. 2008. *Production of Sustainable Alternatives to Petrochemicals and Fuels Using Waste Bread as a Raw Material*, The University of Manchester, Manchester.

Melikoglu, M., C.S.K. Lin, and C. Webb. 2013a. Kinetic studies on the multi-enzyme solution produced via solid state fermentation of waste bread by *Aspergillus awamori*. *Biochemical Engineering Journal* 80:76–82.

Melikoglu, M., C.S.K. Lin, and C. Webb. 2013b. Analysing global food waste problem: Pinpointing the facts and estimating the energy content. *Central European Journal of Engineering* 3 (2):157–164.

Melikoglu, M., C.S.K. Lin, and C. Webb. 2013c. Stepwise optimisation of enzyme production in solid state fermentation of waste bread pieces. *Food and Bioproducts Processing* 91 (4):638–646.

Menon, V., and M. Rao. 2012. Trends in bioconversion of lignocellulose: Biofuels, platform chemicals & biorefinery concept. *Progress in Energy and Combustion Science* 38 (4):522–550.

Merino, S.T., and J. Cherry. 2007. Progress and challenges in enzyme development for biomass utilization. *Advances in Biochemical Engineering Biotechnology* 108:95–120.

Miller, G.L. 1959. Use of dinitrosalicylic acid reagent for determination of reducing sugars. *Analytical Chemistry* 31:426–428.

Moftah, O.A.S., S. Grbavčić, M. Žuža, N. Luković, D. Bezbradica, and Z. Knezevic-Jugovic. 2012. Adding value to the oil cake as a waste from oil processing industry: Production of lipase and protease by *Candida utilis* in solid state fermentation. *Applied Biochemistry and Biotechnology* 166 (2):348–364.

Mohd Yasin, N.H., N.A. Rahman, H.C. Man, M.Z.M. Yusoff, and M.A. Hassan. 2011. Microbial characterization of hydrogen-producing bacteria in fermented food waste at different pH values. *International Journal of Hydrogen Energy* 36:9571–9580.

Molino, A., F. Nanna, Y. Ding, B. Bikson, and G. Braccio. 2013. Biomethane production by anaerobic digestion of organic waste. *Fuel* 103:1003–1009.

Moon, H.C., and I.S. Song. 2011. Enzymatic hydrolysis of foodwaste and methane production using UASB bioreactor. *International Journal of Green Energy* 8 (3):361–371.

Moon, H.C., I.S. Song, J.C. Kim, Y. Shirai, D.H. Lee, J.K. Kim, O.C. Sung, D.H. Kim, K.K. Oh, and Y.S. Cho. 2009. Enzymatic hydrolysis of food waste and ethanol fermentation. *International Journal of Energy Resources* 33 (2):164–172.

Mooney, B.P. 2009. The second green revolution? Production of plant-based biodegradable plastics. *Biochemical Journal* 418 (2):219–232.

Morita, M., and K. Sasaki. 2012. Factors influencing the degradation of garbage in methanogenic bioreactors and impacts on biogas formation. *Applied Microbiology and Biotechnology* 94 (3):575–582.

Mtz. Viturtia, A., J. Mata-Alvarez, F. Cecchi, and G. Fazzini. 1989. Two-phase anaerobic digestion of a mixture of fruit and vegetable wastes. *Biological Wastes* 29 (3):189–199.

Murthy, P.S., M. Madhava Naidu, and P. Srinivas. 2009. Production of α-amylase under solid-state fermentation utilizing coffee waste. *Journal of Chemical Technology Biotechnology* 84 (8):1246–1249.

Muthukumar, M., D. Mohan, and M. Rajendran. 2003. Optimization of mix proportions of mineral aggregates using Box Behnken design of experiments. *Cement and Concrete Composites* 25:751–758.

Nagao, N., N. Tajima, M. Kawai, C. Niwa, N. Kurosawa, T. Matsuyama, F.M. Yusoff, and T. Toda. 2012. Maximum organic loading rate for the single-stage wet anaerobic digestion of food waste. *Bioresource Technology* 118:210–218.

Nasir, I.M., T.I.M. Ghazi, and R. Omar. 2012. Production of biogas from solid organic wastes through anaerobic digestion: A review. *Applied Microbiology and Biotechnology* 95 (2):321–329.

National Environment Agency (Singapore). 2013. http://www.nea.gov.sg/. National Environmental Agency 2013 [cited 3 February 2013].

Neste-oil. 2012. *Future Renewable Raw Materials* 2014 [cited 29 August 2014]. Annual report available from: https://www.neste.com/en/corporate-info/news-media/material-uploads/annual-reports-0.

Ngoc, U.N., and H. Schnitzer. 2009. Sustainable solutions for solid waste management in Southeast Asian countries. *Waste Management* 29:1982–1995.

Noor, Z.Z., R.O. Yusuf, A.H. Abba, M.A. Abu Hassan, and M.F. Mohd Din. 2013. An overview for energy recovery from municipal solid wastes (MSW) in Malaysia scenario. *Renewable and Sustainable Energy Reviews* 20:378–384.

Norouzian, D., A. Akbarzadeh, J.M. Scharer, and M. Moo Young. 2006. Fungal glucoamylases. *Biotechnology Advances* 24 (1):80–85.

Novikova, L.N., J. Pettersson, M. Brohlin, M. Wiberg, and L.N. Novikov. 2008. Biodegradable poly-b-hydroxybutyrate scaffold seeded with Schwann cells to promote spinal cord repair. *Biomaterials* 29 (9):1198–1206.

Oberoi, H.S., P.V. Vadlani, A. Nanjundaswamy, S. Bansal, S. Singh, S. Kaur, and N. Babbar. 2011a. Enhanced ethanol production from Kinnow mandarin (*Citrus reticulata*) waste via a statistically optimized simultaneous saccharification and fermentation process. *Bioresource Technology* 102:1593–1601.

Oberoi, H.S., P.V. Vadlani, L. Saida, S. Bansal, and J.D. Hughes. 2011b. Ethanol production from banana peels using statistically optimized simultaneous saccharification and fermentation process. *Waste Management* 31:1576–1584.

OECD. 2007. *Municipal Waste Generation.* OECD Publishing, https://data.oecd.org/waste/municipal-waste.htm (accessed June 2017).

Ohkouchi, Y., and Y. Inoue. 2006. Direct production of L(+)-lactic acid from starch and food wastes using *Lactobacillus manihotivorans* LMG18011. *Bioresource Technology* 97:1554–1562.

Ohkouchi, Y., and Y. Inoue. 2007. Impact of chemical components of organic wastes on l(+)-lactic acid production. *Bioresource Technology* 98 (3):546–553.

Omar, F.N., N.A. Rahman, H.S. Hafid, T. Mumtaz, P.L. Yee, and M.A. Hassan. 2011. Utilization of kitchen waste for the production of green thermoplastic polyhydroxybutyrate (PHB) by *Cupriavidus necator* CCGUG 52238. *African Journal of Microbiology Research* 5 (19):2873–2879.

Omar, F.N., N.A. Rahman, H.S. Hafid, P.L. Yee, and M.A. Hassan. 2009. Separation and recovery of organic acids from fermented kitchen waste by an integrated process. *African Journal of Biotechnology* 8 (21):5807–5813.

Othman, S.N., Z.Z. Noor, A.H. Abba, R.O. Yusuf, and M.A. Abu Hassan. 2013. Review on life cycle assessment of integrated solid waste management in some Asian countries. *Journal of Cleaner Production* 41:251–262.

Pan, J., R. Zhang, H.M. El-Mashad, H. Sun, and Y. Ying. 2008. Effect of food to microorganism ratio on biohydrogen production from food waste via anaerobic fermentation. *International Journal of Hydrogen Energy* 33 (23):6968–6975.

Pandey, A. 1991. Effect of particle size of substrate on enzyme production in solid-state fermentation. *Bioresource Technology* 37:169–172.

Pandey, A., P. Nigam, C.R. Soccol, V.T. Soccol, D. Singh, and R. Mohan. 2000a. Advances in microbial amylases. *Biotechnology and Applied Biochemistry* 31 (2):135–152.

Pandey, A., C.R. Soccol, P. Nigam, and V.T. Soccol. 2000b. Biotechnological potential of agro-industrial residues. I: Sugarcane bagasse. *Bioresource Technology* 74 (1):69–80.

Papanikolaou, S., and G. Aggelis. 2003. Selective uptake of fatty acids by the yeast *Yarrowia lipolytica*. *European Journal of Lipid Science and Technology* 105:651–655.

Papanikolaou, S., and G. Aggelis. 2011. Lipids of oleaginous yeasts. Part I: Biochemistry of single cell oil production. *European Journal of Lipid Science and Technology* 113 (8):1031–1051.

Papanikolaou, S., I. Chevalot, M. Galiotou-Panayotou, M. Komaitis, and G. Aggelis. 2007. Industrial derivative of tallow: A promising renewable substrate for microbial lipid, single-cell protein and lipase production by *Yarrowia lipolytica*. *Electronic Journal of Biotechnology* 10 (3):425–435.

Papanikolaou, S., I. Chevalot, M. Komaitis, I. Marc, and G. Aggelis. 2002. Single cell oil production by *Yarrowia lipolytica* growing on an industrial derivative of animal fat in batch cultures. *Applied Microbiology and Biotechnology* 58 (3):308–312.

Papanikolaou, S., A. Dimou, S. Fakas, P. Diamantopoulou, A. Philippoussis, M. Galiotou-Panayotou, and G. Aggelis. 2011. Biotechnological conversion of waste cooking olive oil into lipid-rich biomass using *Aspergillus* and *Penicillium* strains. *Journal of Applied Microbiology* 110 (5):1138–1150.

Papanikolaou, S., L. Muniglia, I. Chevalot, G. Aggelis, and I. Marc. 2003. Accumulation of a cocoa-butter-like lipid by yarrowia lipolytica cultivated on agro-industrial residues. *Current Microbiology* 46 (2):124–130.

Parawira, W., M. Murto, J.S. Read, and B. Mattiasson. 2005. Profile of hydrolases and bio-gas production during two-stage mesophilic anaerobic digestion of solid potato waste. *Process Biochemistry* 40:2945–2952.

Park, Y.J., F. Hong, J.H. Cheon, T. Hidaka, and H. Tsuno. 2008. Comparison of thermophilic anaerobic digestion characteristics between single-phase and two-phase systems for kitchen garbage treatment. *Journal of Bioscience and Bioengineering* 105 (1):48–54.

Patel, S.K.S., P. Kumar, and V.C. Kalia. 2012. Enhancing biological hydrogen production through complementary microbial metabolisms. *International Journal of Hydrogen Energy* 37 (14):10590–10603.

Pedrolli, D.B., E. Gomes, R. Monti, and E.C. Carmona. 2008. Studies on productivity and characterization of polygalacturonase from *Aspergillus giganteus* submerged culture using citrus pectin and orange waste. *Applied Biochemistry and Biotechnology* 144 (2):191–200.

Pedrolli, D.B., A.C. Monteiro, E. Gomes, and E.C. Carmona. 2009. Pectin and pectinases: Production, characterization and industrial application of microbial pectinolytic enzymes. *Open Biotechnology Journal* 3:9–18.

Pfaltzgraff, L.A., M. De bruyn, E.C. Cooper, V. Budarin, and J.H. Clark. 2013. Food waste biomass: A resource for high-value chemicals. *Green Chemistry* 15:307–314.

Pleissner, D., T.H. Kwan, and C.S.K. Lin. 2014. Fungal hydrolysis in submerged fermentation for food waste treatment and fermentation feedstock preparation. *Bioresource Technology* 158:48–54.

Pleissner, D., W.C. Lam, Z. Sun, and C.S.K. Lin. 2013. Food waste as nutrient source in heterotrophic microalgae cultivation. *Bioresource Technology* 137:139–146, doi: http://dx.doi.org/10.1016/j.biortech.2013.03.088.

Potumarthi, R., S. Ch, and A. Jetty. 2007. Alkaline protease production by submerged fermentation in stirred tank reactor using *Bacillus licheniformis* NCIM-2042: Effect of aeration and agitation regimes. *Biochemical Engineering Journal* 34:185–192.

Prakasham, R.S., C. Subba Rao, R. Sreenivas Rao, and P.N. Sarma. 2005. Alkaline protease production by an isolated *Bacillus circulans* under solid-state fermentation using agroindustrial waste: Process parameters optimization. *Biotechnology Progress* 21 (5):1380–1388.

Punrattanasin, W., A.A. Randall, and C.W. Randall. 2006. Aerobic production of activated sludge polyhydroxyalkanoates from nutrient deficient wastewaters. *Water Science and Technology* 54 (8):1–8.

Quiroga, G., L. Castrillón, Y. Fernández-Nava, E. Marañón, L. Negral, J. Rodríguez-Iglesias, and P. Ormaechea. 2014. Effect of ultrasound pre-treatment in the anaerobic co-digestion of cattle manure with food waste and sludge. *Bioresoure Technology* 154:74–79.

Ramos, C., G. Buitron, I. Moreno-Andrade, and R. Chamy. 2012. Effect of the initial total solids concentration and initial pH on the bio-hydrogen production from cafeteria food waste. *International Journal of Hydrogen Energy* 37:13288–13295.

Ramrakhiani, L., and S. Chand. 2011. Recent progress on phospholipases: Different sources, assay methods, industrial potential and pathogenicity. *Applied Biochemistry and Biotechnology* 164:991–1022.

Rao, M.S., and S.P. Singh. 2004. Bioenergy conversion studies of organic fraction of MSW: Kinetic studies and gas yield–organic loading relationships for process optimisation. *Bioresource Technology* 95:173–185.

Rao, M.S., and W.F. Stevens. 2006. Fermentation of shrimp biowaste under different salt concentrations with amylolytic and non-amylolytic *Lactobacillus* strains for chitin production. *Food Technology and Biotechnology* 44 (1):83–87.

Ratledge, C. 1991. Microorganisms for lipids. *Acta Biotechnologica* 11 (5):429–438.

Ratledge, C., and J.P. Wynn. 2002. The biochemistry and molecular biology of lipid accumulation in oleaginous microorganisms. *Advances in Applied Microbiology* 51:1–51.

Rehman, S., H.N. Bhatti, I.A. Bhatti, and M. Asgher. 2011. Optimization of process parameters for enhanced production of lipase by *Penicillium notatum* using agricultural wastes. *African Journal of Biotechnology* 10 (84):19580–19589.

Rhu, D.H., W.H. Lee, J.Y. Kim, and E. Choi. 2003. Polyhydroxyalkanoate (PHA) production from waste. *Water Science and Technology* 48 (8): 221–228.

Rivas, B., A. Torrado, P. Torre, A. Converti, and J.M. Domínguez. 2008. Submerged citric acid fermentation on orange peel autohydrolysate. *Journal of Agricultural and Food Chemistry* 56 (7):2380–2387.

Rivas-Cantu, R.C., K.D. Jones, and P.L. Mills. 2013. A citrus waste-based biorefinery as a source of renewable energy: Technical advances and analysis of engineering challenges. *Waste Management and Research* 31 (4):413–420.

Rodrigues, C., L.P.S. Vanderberghe, and C.R. Soccol. 2009. Improvement in citric acid production in solid-state fermentation by *Aspergillus niger* LPB BC mutant using citric pulp. *Applied Biochemistry and Biotechnology* 158:72–87.

Ruiz, H.A., R.M. Rodriguez-Jasso, R. Rodriguez, J.C. Contreras-Esquivel, and C.N. Aguilar. 2012. Pectinase production from lemon peel pomace as support and carbon source in solid state fermentation column-tray bioreactor. *Biochemical Engineering Journal* 65:90–95.

Rusendi, D., and J.D. Sheppard. 1995. Hydrolysis of potato processing waste for the production of poly-β-hydroxybutyrate. *Bioresource Technology* 54 (2):191–196.

Sakai, K., and Y. Ezaki. 2006. Open L-lactic acid fermentation of food refuse using thermophilic *Bacillus coagulans* and fluorescence in situ hybridization analysis of microflora. *Journal of Bioscience and Bioengineering* 101 (6):457–463.

Sakai, K., N. Fujii, and E. Chukeatirote. 2006. Racemization of L-lactic acid in pH-swing open fermentation of kitchen refuse by selective proliferation of *Lactobacillus plantarum*. *Journal of Bioscience and Bioengineering* 102 (3):227–232.

Sakai, K., M. Mori, A. Fujii, Y. Iwami, E. Chuleatiote, and Y. Shirai. 2004a. Fluorescent in situ hybridization analysis of open lactic acid fermentation of kitchen refuse using rRNA targeted oligonucleotide probes. *Journal of Bioscience and Bioengineering* 98 (1):48–56.

Sakai, K., Y. Murata, H. Yamazumi, Y. Tau, M. Mori, M. Moriguchi, and Y. Shirai. 2000. Selective proliferation of lactic acid bacteria and accumulation of lactic acid during open fermentation of kitchen refuse with intermittent pH adjustment. *Food Science and Technology Research* 6 (2):140–145.

Sakai, K., M. Taniguchi, S. Miura, H. Ohara, T. Matsumoto, and Y. Shirai. 2004b. Making plastics from garbage: A novel process for poly-L-lactate production from municipal food waste. *Journal of Industrial Ecology* 7 (3–4):63–74.

Sakai, K., and T. Yamanami. 2006. Thermotolerant *Bacillus licheniformis* TY7 produces optically active L-lactic acid from kitchen refuse under open condition. *Journal of Bioscience and Bioengineering* 102 (2):132–134.

Sanders, J., E. Scott, R. Weusthuis, and H. Mooibroek. 2007. Bio-refinery as the bio-inspired process to bulk chemicals. *Macromolecular Bioscience* 7 (2):105–117.

Saravanan, P., R. Muthuvelayudham, and T. Viruthagiri. 2012. Application of statistical design for the production of cellulase by *Trichoderma reesei* using mango peel. *Enzyme Research* 2012:157643–157649.

Satoh, H., Y. Iwamoto, T. Mino, and T. Matsuo. 1998. Activated sludge as a possible source of biodegradable plastic. *Water Science and Technology* 38 (2):103–109.

Saxena, R.K., W.S. Davidson, A. Sheoran, and B. Giri. 2003. Purification and characterization of an alkaline thermostable lipase from *Aspergillus carneus*. *Process Biochemistry* 39 (2):239–247.

Schmidt, C.G., and E.B. Furlong. 2012. Effect of particle size and ammonium sulfate concentration on rice bran fermentation with the fungus *Rhizopus oryzae*. *Bioresource Technology* 123:36–41.

Sellami, M., S. Kedachi, F. Frikha, N. Miled, and F. Ben Rebah. 2013. Optimization of marine waste based-growth media for microbial lipase production using mixture design methodology. *Environmental Technology* 34 (15):2259–2266.

Seng, B.B. 2010. Municipal solid waste management in Phnom Penh, capital city of Cambodia. *Waste Management and Research* 29:491–500.

Seo, Y.H., I. Lee, S.H. Jeon, and J. Han. 2014. Efficient conversion from cheese whey to lipid using *Cryptococcus curvatus*. *Biochemical Engineering Journal* 90:149–153.

Sharma, N., K.L. Kalra, H.S. Oberoi, and S. Bansal. 2007. Optimization of fermentation parameters for production of ethanol from kinnow waste and banana peels by simultaneous saccharifi cation and fermentation. *Indian Journal of Microbiology* 47:310–316.

Shen, F., R. Liu, and T. Wang. 2009. Effects of temperature, pH, agitation and particles stuffing rate on fermentation of sorghum stalk juice to ethanol. *Energy Sources, Part A: Recovery, Utilization and Environmental Effects* 31 (8):646–656.

Shin, H., and J. Youn. 2005. Conversion of food waste into hydrogen by thermophilic acidogenesis. *Biodegradation* 16 (1):33–44.

Shin, H.S., J.-S. Youn, and S.H. Kim. 2004. Hydrogen production from food waste in anaerobic mesophilic and thermophilic acidogenesis. *International Journal of Hydrogen Energy* 29:1355–1363.

Shojaosadati, S.A., and V. Babaeipour. 2002. Citric acid production from apple pomace in multi-layer packed bed solid-state bioreactor. *Process Biochemistry* 37 (8):909–914.

Show, K.Y., D.J. Lee, J.H. Tay, C.Y. Lin, and J.S. Chang. 2012. Biohydrogen production: Current perspectives and the way forward. *International Journal of Hydrogen Energy* 37:15616–15631.

Shukla, J., and R. Kar. 2006. Potato peel as a solid state substrate for thermostable alpha amylase production by thermophilic *Bacillus* isolates. *World Journal of Microbiology and Biotechnology* 22 (5):417–422.

Smith, A.D., M. Landoll, M. Falls, and M.T. Holtzapple. 2010. Chemical production from lignocellulosic biomass: Thermochemical, sugar and carboxylate platforms. In *Bioalcohol Production* (Waldron, K. ed.), Woodhead Publishing Limited, Cambridge, UK.

Solaiman, D.K.Y., R.D. Ashby, T.A. Foglia, and W.N. Marmer. 2006. Conversion of agricultural feedstock and coproducts into poly(hydroxyalkanoates). *Applied Microbiology Biotechnology* 71 (6):783–789.

Soni, S.K., A. Kaur, and J.K. Gupta. 2003. A solid-state fermentation based bacterial alpha-amylase, fungal glucoamylase system, its suitability for hydrolysis of wheat starch. *Process Biochemistry* 39:158–192.

Souissi, N., Y. Ellouz-Triki, A. Bougatef, M. Blibech, and M. Nasri. 2008. Preparation and use of media for protease-producing bacterial strains based on by-products from cuttlefish (*Sepia officinalis*) and wastewaters from marine-products processing factories. *Microbiological Research* 163 (4):473–480.

ST1. 2013. St1 Biofuels bioethanol production (2012), 22.04.2013 2013 [cited 2 April 2013].

Steinbuchel, A., and B. Fuchtenbusch. 1998. Bacterial and other biological systems for polyester production. *Trends in Biotechnology* 16 (10):419–427.

Sun, H., X. Ge, Z. Hao, and M. Peng. 2010. Cellulase production by *Trichoderma* sp. on apple pomace under solid state fermentation. *African Journal of Biotechnology* 9 (2):163–166.

Sun, H.Y., J. Li, P. Zhao, and M. Peng. 2011. Banana peel: A novel substrate for cellulase production under solid-state fermentation. *African Journal of Biotechnology* 10 (77):17887–17890.

Sun, Z., M. Li, Q. Qi, C. Gao, and C.S.K. Lin. 2014. Mixed food waste as renewable feedstock in succinic acid fermentation. *Applied Biochemistry and Biotechnology* 174:1822–1833.

Takata, M., K. Fukushima, N. Kino-Kimata, N. Nagao, C. Niwa, and T. Toda. 2012. The effects of recycling loops in food waste management in Japan: Based on the environmental and economic evaluation of food recycling. *Science of the Total Environment* 432:309–317.

Tang, Y.Q., Y. Koike, K. Liu, M.Z. An, S. Morimura, X.L. Wu, and K. Kida. 2008. Ethanol production from kitchen waste using the flocculating yeast *Saccharomyces cerevisiae* strain KF-7. *Biomass and Bioenergy* 32 (11):1037–1045.

Tao, F., J.Y. Miao, G.Y. Shi, and K.C. Zhang. 2005. Ethanol fermentation by an acid-tolerant *Zymomonas mobilis* under non-sterilized condition. *Process Biochemistry* 40:183–187.

Taylor, P. 2010. Biosuccinic acid ready for take off? *Chemistry World* 7 (2):16–17.

Teeri, T.T. 1997. Crystalline cellulose degradation: New insight into the function of cellobiohydrolases. *Trends in Biotechnology* 15:160–167.

Thomas, L., C. Larroche, and A. Pandey. 2013. Current developments in solid-state fermentation. *Biochemical Engineering Journal* 81:146–161.

Thomsen, A.B., C. Medina, and B.K. Ahring. 2003. Biotechnology in ethanol production. In *New and Emerging Bioenergy Technologies*, edited by H. Larsen, Kossman, J., Petersen, L.S. Denmark: 2. Riso National Laboratory.

Tomasik, P., and D. Horton. 2012. Enzymatic conversions of starch. *Advances in Carbohydrate Chemistry and Biochemistry* 68, 59–436.

Torrado, A.M., S. Cortés, J.M. Salgado, B. Max, N. Rodríguez, B.P. Bibbins, A. Converti, and J.M. Domínguez. 2011. Citric acid production from orange peel wastes by solid-state fermentation. *Brazilian Journal of Microbiology* 42 (1):394–409.

Toscano, L., G. Montero, M. Stoytcheva, V. Gochev, L. Cervantes, H. Campbell, R. Zlatev, B. Valdez, C. Pérez, and M. Gil-Samaniego. 2013. Lipase production through solid-state fermentation using agro-industrial residues as substrates and newly isolated fungal strains. *Biotechnology and Biotechnological Equipment* 27 (5):4074–4077.

Trzcinski, A.P., and D.C. Stuckey. 2011. Parameters affecting the stability of the digestate from a two-stage anaerobic process treating the organic fraction of municipal solid waste. *Waste Management* 31 (7):1480–1487.

Trzcinski, A.P., and D.C. Stuckey. 2012. Determination of the hydrolysis constant in the bio-chemical methane potential test of municipal solid waste. *Environmental Engineering Science* 29 (9):848–854.

Tubb, R.S. 1986. Amylolytic yeasts for commercial applications. *Trends in Biotechnology* 4 (4):98–104.

Tuck, C.O., E. Pérez, I.T. Horváth, R.A. Sheldon, and M. Poliakoff. 2012. Valorization of biomass: Deriving more value from waste. *Science* 337:695–699.

Uçkun, E.K., A.P. Trzcinski, and Y. Liu. 2015a. Platform chemicals production from food wastes using a biorefinery concept. *Journal of Chemical Technology and Biotechnology* 90 (8):1364–1379.

Uçkun, K.E., O. Akpinar, and U. Bakir. 2013a. Improvement of enzymatic xylooligosac-charides production by the co-utilization of xylans from different origins. *Food and Bioproducts Processing* 91 (4):565–574.

Uçkun, K.E., and Y. Liu. 2015. Bioethanol production from mixed food waste by an effective enzymatic pretreatment. *Fuel* 159:463–469.

Uçkun, K.E., A. Salakkam, A. Trzcinski, U. Bakir, and C. Webb. 2012. Enhancing the value of nitrogen from rapeseed meal for microbial oil production. *Enzyme and Microbial Technology* 50:337–342.

Uçkun, K.E., A.P. Trzcinski, and Y. Liu. 2015b. Enhancing the hydrolysis and methane pro-duction potential of mixed food wastes by a cost-effective enzymatic pretreatment. *Bioresource Technology* 183:47–52.

Uçkun, K.E., A.P. Trzcinski, Y. Liu, and W.J. Ng. 2014a. Enzyme production from food wastes using a biorefinery concept: A review. *Waste and Biomass Valorization* 5 (6):903–917.

Uçkun, K.E., A.P. Trzcinski, and Y. Liu. 2014b. Glucoamylase production from food waste by solid state fermentation and its evaluation in the hydrolysis of domestic food waste. *Biofuel Research J.* 3:98–105.

Uçkun, K.E., A.P. Trzcinski, W.J. Ng, and Y. Liu. 2014c. Bioconversion of food waste to energy: A review. *Fuel* 134:389–399.

Uçkun, K.E., A. Trzcinski, and C. Webb. 2013b. Microbial oil produced from biodiesel by-products could enhance overall production. *Bioresource Technology* 129:650–654.

Umsza-Guez, M.A., A.B. Díaz, I. de Ory, A. Blandino, E. Gomes, and I. Caro. 2011. Xylanase production by *Aspergillus awamori* under solid state fermentation conditions on tomato pomace. *Brazilian Journal of Microbiology* 42 (4):1585–1597.

Uncu, O.N., and D. Cekmecelioglu. 2011. Cost-effective approach to ethanol production and optimization by response surface methodology. *Waste Management* 31 (4):636–643.

Vamvakaki, A.N., I. Kandarakis, S. Kaminarides, M. Komaitis, and S. Papanikolaou. 2010. Cheese whey as a renewable substrate for microbial lipid and biomass production by *Zygomycetes*. *Engineering in Life Sciences* 10 (4):348–360.

Vaseghi, Z., G.D. Najafpour, S. Mohseni, and S. Mahjoub. 2013. Production of active lipase by *Rhizopus oryzae* from sugarcane bagasse: Solid state fermentation in a tray bioreac-tor. *International Journal of Food Science and Technology* 48 (2):283–289.

Vavouraki, A.I., E.M. Angelis, and M. Kornaros. 2014. Optimization of thermo-chemical hydrolysis of kitchen wastes. *Waste Management* 34 (1):167–173.

Venkateswar Reddy, M., and S. Venkata Mohan. 2012. Influence of aerobic and anoxic micro-environments on polyhydroxyalkanoates (PHA) production from food waste and acido-genic effluents using aerobic consortia. *Bioresource Technology* 103 (1):313–321.

Wang, F.S., and H.T. Lin. 2010. Fuzzy optimization of continuous fermentations with cell recycling for ethanol production. *Industrial and Engineering Chemistry Research* 49 (5):2306–2311.

Wang, G., Z. Wang, Y. Zhang, and Y. Zhang. 2012. Cloning and expression of amyE gene from *Bacillus subtilis* in *Zymomonas mobilis* and direct production of ethanol from soluble starch. *Biotechnology and Bioprocess Engineering* 17 (4):780–786.

Wang, H., J. Wang, Z. Fang, X. Wang, and H. Bu. 2010a. Enhanced bio-hydrogen production by anaerobic fermentation of apple pomace with enzyme hydrolysis. *International Journal of Hydrogen Energy* 35:8303–8309.

Wang, M., Z. Xu, T. Qiu, X. Sun, M. Han, and X. Wang. 2010b. Kinetics of lactic acid fermentation on food waste by *Lactobacillus bulgaricus*. *Advanced Materials Research* 113 (116):1235–1238.

Wang, Q.H., Y.Y. Liu, and H.Z. Ma. 2010c. On-site production of crude glucoamylase for kitchen waste hydrolysis. *Waste Management and Research* 28:539–544.

Wang, Q., H. Ma, W. Xu, L. Gong, W. Zhang, and D. Zou. 2008a. Ethanol production from kitchen garbage using response surface methodology. *Biochemical Engineering Journal* 39:604–610.

Wang, Q., J. Narita, W. Xie, Y. Ohsumi, K. Kusano, Y. Shirai, and H.I. Ogawa. 2002. Effects of anaerobic/aerobic incubation and storage temperature on preservation and deodorization of kitchen garbage. *Bioresource Technology* 84 (3):213–220.

Wang, Q., X. Wang, X. Wang, and H. Ma. 2008b. Glucoamylase production from food waste by *Aspergillus niger* under submerged fermentation. *Process Biochemistry* 43 (3):280–286.

Wang, Q., X. Wang, X. Wang, H. Ma, and N. Ren. 2005a. Bioconversion of kitchen garbage to lactic acid by two wild strains of *Lactobacillus* species. *Journal of Environmental Science and Health—Part A Toxic/Hazardous Substances and Environmental Engineering* 40 (10):1951–1962.

Wang, Q., H. Ma, W. Xu, L. Gong, W. Zhang, and D. Zou. 2008c. Ethanol production from kitchen garbage using response surface methodology. *Biochemical Engineering Journal* 39 (3):604–610.

Wang, R., L.C. Godoy, S.M. Shaarani, M. Melikoglu, A. Koutinas, and C. Webb. 2009a. Improving wheat flour hydrolysis by an enzyme mixture from solid state fungal fermentation. *Enzyme and Microbial Technology* 44 (4):223–228.

Wang, R., Y. Ji, M. Melikoglu, A. Koutinas, and C. Webb. 2007. Optimization of innovative ethanol production from wheat by response surface methodology. *Process Safety and Environmental Protection* 85 (5B):404–412.

Wang, X.M. 2011. Effect of different fermentation parameters on lactic acid production from kitchen waste by *Lactobacillus* TY50. *Chemical and Biochemical Engineering Quarterly* 25 (4):433–438.

Wang, X., Q. Wang, X. Wang, and H. Ma. 2011. Effect of different fermentation parameters on lactic acid production from kitchen waste by *Lactobacillus* TY50. *Chemical and Biochemical Engineering Quarterly* 25 (4):433–438.

Wang, X., and Y.C. Zhao. 2009. A bench scale study of fermentative hydrogen and methane production from food waste in integrated two-stage process. *International Journal of Hydrogen Energy* 34:245–254.

Wang, X.M., Q.H. Wang, N.Q. Ren, and X.Q. Wang. 2005b. Lactic acid production from kitchen waste with a newly characterized strain of *Lactobacillus plantarum*. *Chemical and Biochemical Engineering Quarterly* 19 (4):383–389.

Wang, X.Q., Q.H. Wang, H.Z. Ma, and W. Yin. 2009b. Lactic acid fermentation of food waste using integrated glucoamylase production *Journal of Chemical Technology and Biotechnology* 84 (1):139–143.

Ward, O. P., and A. Singh. 2005. Omega-3/6 fatty acids: Alternative sources of production. *Process Biochemistry* 40 (12):3627–3652.

Wee, Y., J. Kim, and H. Ryu. 2006. Biotechnological production of lactic acid and its recent applications. *Food Technology and Biotechnology* (2):163–172.

Weil, J., P. Westgate, K. Kohlmann, and M.R. Ladisch. 1994. Cellulose pretreatments of lignocellulosic substrates. *Enzyme and Microbial Technology* 16 (11):1002–1004.

Xu, Y., R. Wang, A.A. Koutinas, and C. Webb. 2010. Microbial biodegradable plastic production from a wheat-based biorefining strategy. *Process Biochemistry* 45 (2):153–163.

Yaakob, Z., M. Mohammad, M. Alherbawi, Z. Alam, and K. Sopian. 2013. Overview of the production of biodiesel from waste cooking oil. *Renewable and Sustainable Energy Reviews* 18:184–193.

Yadav, A.K., A.B. Chaudhari, and R.M. Kothari. 2011. Bioconversion of renewable resources into lactic acid: An industrial view. *Critical Reviews in Biotechnology* 31 (1):1–19.

Yamane, T. 1993. Yield of poly-(o)-3-hydroxybutyrate from various carbon sources: A theoretical study. *Biotechnology and Bioengineering* 41 (1):165–170.

Yan, S., X. Chen, J. Wu, and P. Wang. 2012a. Ethanol production from concentrated food waste hydrolysates with yeast cells immobilized on corn stalk. *Applied Microbiology Biotechnology* 94 (3):829–838.

Yan, S., J. Yao, L. Yao, Z. Zhi, X. Chen, and J. Wu. 2012b. Fed batch enzymatic saccharification of food waste improves the sugar concentration in the hydrolysates and eventually the ethanol fermentation by *Saccharomyces cerevisiae* H058. *Brazilian Archives of Biology and Technology* 55:183–192.

Yang, S.Y., K.S. Ji, Y.H. Baik, W.S. Kwak, and T.A. McCaskey. 2006. Lactic acid fermentation of food waste for swine feed. *Bioresource Technology* 97 (15):1858–1864.

Ye, Z., Y. Zheng, Y. Li, and W. Cai. 2008a. Use of starter culture of *Lactobacillus plantarum* BP04 in the preservation of dining-hall food waste. *World Journal of Microbiology and Biotechnology* 24 (10):2249–2256.

Ye, Z.L., M. Lu, Y. Zheng, Y.H. Li, and W.M Cai. 2008b. Lactic acid production from dining-hall food waste by *Lactobacillus plantarum* using response surface methodology. *Journal of Chemical Technology and Biotechnology* 83 (11):1541–1550.

Yin, Y., Y-J. Liu, S-J. Meng, E. Uçkun, and Y. Liu. 2016. Enzymatic pretreatment of activated sludge, food waste and their mixture for enhanced bioenergy recovery and waste volume reduction via anaerobic digestion. *Applied Energy* 179:1131–1137.

Youn, J., and H. Shin. 2005. Comparative performance between temperature-phased and conventional mesophilic two-phased processes in terms of anaerobically produced bioenergy from food waste. *Waste Management and Research* 23 (1):32–38.

Yu, P.H., H. Chua, A.L. Huang, and K.P. Ho. 1999. Conversion of industrial food wastes by *Alcaligenes latus* into polyhydroxyalkanoates. *Applied Biochemistry and Biotechnology—Part A Enzyme Engineering and Biotechnology* 77–79, 445–454.

Zeikus, J.G., M.K. Jain, and P. Elankovan. 1999. Biotechnology of succinic acid production and markets for derived industrial products. *Applied Microbiology and Biotechnology* 51 (5):545–552.

Zhang, A.Y., Z. Sun, C.C.J. Leung, W. Han, K.Y. Lau, M. Li, and C.S.K. Lin. 2013a. Valorisation of bakery waste for succinic acid production. *Green Chemistry* 15 (3):690–695.

Zhang, B., P.J. He, N.F. Ye, and L.M. Shao. 2008. Enhanced isomer purity of lactic acid from the non-sterile fermentation of kitchen wastes. *Bioresource Technology* 99:855–862.

Zhang, B., Z. He, L. Zhang, J. Xu, H. Shi, and W. Cai. 2005. Anaerobic digestion of kitchen wastes in a single-phased anaerobic sequencing batch reactor (ASBR) with gas-phased absorb of CO2. *Journal of Environmental Sciences* 17 (2):249–255.

Zhang, C., H. Su, Baeyens, and T. Tan. 2014. Reviewing the anaerobic digestion of food waste for biogas production. *Renewable and Sustainable Energy Reviews* 38:383–392.

Zhang, C., G. Xiao, L. Peng, H. Su, and T. Tan. 2013b. The anaerobic co-digestion of food waste and cattle manure. *Bioresource Technology* 129:170–176.

Zhang, G., W.T. French, R. Hernandez, J. Hall, D. Sparks, and W.E. Holmes. 2011. Microbial lipid production as biodiesel feedstock from N-acetylglucosamine by oleaginous microorganisms. *Journal of Chemical Technology and Biotechnology* 86 (5):642–650.

Zhang, L., and D. Jahng. 2012. Long-term anaerobic digestion of food waste stabilized by trace elements. *Waste Management* 32 (8):1509–1515.

Zhang, M., P. Shukla, M. Ayyachamy, K. Permaul, and S. Singh. 2010. Improved bioethanol production through simultaneous saccharification and fermentation of lignocellulosic agricultural wastes by *Kluyveromyces marxianus* 6556. *World Journal of Microbiology and Biotechnology* 26 (6):1041–1046.

Zhang, M., H. Wu, and H. Chen. 2013c. Coupling of polyhydroxyalkanoate production with volatile fatty acid from food wastes and excess sludge. *Process Safety and Environmental Protection* 92 (2):171–178.

Zhang, R., H.M. El-Mashad, K. Hartman, F. Wang, G. Liu, C. Choate, and P. Gamble. 2007. Characterization of food waste as feedstock for anaerobic digestion. *Bioresource Technology* 98 (4):929–935.

Zilly, A., G.C. dos Santos Bazanella, C.V. Helm, C.A. Vaz Araújo, C.G.M. de Souza, A. Bracht, and R.M. Peralta. 2012. Solid-state bioconversion of passion fruit waste by white-rot fungi for production of oxidative and hydrolytic enzymes. *Food Bioprocess Technology* 5 (5):1573–1580.

# Index